DEVELOPMENTS IN SPEECH SYNTHESIS

DEVELOPMENTS IN SPEECH SYNTHESIS

Mark Tatham

Department of Language and Linguistics, University of Essex, UK

Katherine Morton

Formerly University of Essex, UK

John Wiley & Sons, Ltd

This publication is designed to provide accurate and authoritative information in regard to the subject matter covered. It is sold on the understanding that the Publisher is not engaged in rendering professional services. If professional advice or other expert assistance is required, the services of a competent professional should be sought.

Other Wiley Editorial Offices

John Wiley & Sons Inc., 111 River Street, Hoboken, NJ 07030, USA

Jossey-Bass, 989 Market Street, San Francisco, CA 94103-1741, USA

Wiley-VCH Verlag GmbH, Boschstr. 12, D-69469 Weinheim, Germany

John Wiley & Sons Australia Ltd, 33 Park Road, Milton, Queensland 4064, Australia

John Wiley & Sons (Asia) Pte Ltd, 2 Clementi Loop #02-01, Jin Xing Distripark, Singapore 129809

John Wiley & Sons Canada Ltd, 22 Worcester Road, Etobicoke, Ontario, Canada M9W 1L1

Wiley also publishes its books in a variety of electronic formats. Some content that appears in print may not be available in electronic books.

British Library Cataloguing in Publication Data

A catalogue record for this book is available from the British Library

ISBN 0-470-85538-X (HB)

Typeset in 10/12pt Times by Graphicraft, Limited, Hong Kong, China.
Printed and bound in Great Britain by Antony Rowe Ltd, Chippenham, Wiltshire.
This book is printed on acid-free paper responsibly manufactured from sustainable forestry in which at least two trees are planted for each one used for paper production.

Contents

Acknowledgements

We should like to acknowledge British Telecom Research Labs, IBM-UK Research, the UK Department of Trade and Industry, and the UK Engineering and Physical Sciences Research Council (formerly the Science Research Council) for early support for our work on the integration of cognitive and physical approaches in the speech synthesis environment. Thanks in general are due to colleagues in the area of computer modelling of speech production and perception, and in particular to Eric Lewis (Department of Computer Science, University of Bristol).

Introduction

How Good is Synthetic Speech?

In this book we aim to introduce and discuss some current developments in speech synthesis, particularly at the higher level, which focus on some specific issues. We shall see how these issues have arisen and look at possible ways in which they might be dealt with. One of our objectives will be to suggest that a more unified approach to synthesis than we have at the present time may result in overall improvement to synthesis systems.

In the early days of speech synthesis research the obvious focus of attention was *intelligibility*–whether or not the synthesiser's output could be understood by a human listener (Keller 1994; Holmes and Holmes 2001). Various methods of evaluation were developed which often involved comparison between different systems. Interestingly, intelligibility was almost always taken to mean *segmental intelligibility*–that is, whether or not the speech segments which make up words were sufficiently well rendered to enable those words to be correctly recognised. Usually tests for intelligibility were not performed on systems engaged in dialogue with humans–the test environment involved listeners evaluating a synthesiser just speaking to them with no interaction in the form of dialogue. The point here is that intelligibility varies with context, and a dialogue simulation would today be a much more appropriate test environment for intelligibility.

It is essential for synthesisers to move away from the basic requirements of minimally converting text-to-speech–see Dutoit (1997) for a comprehensive overview–to systems which place more emphasis on naturalness of speech production. This will mean that the earlier synthesis model will necessarily become inadequate as the focus shifts from the reading task *per se* to the quality of the synthetic voice.

Improvements Beyond Intelligibility

Although for several years synthetic speech has been fully intelligible from a segmental perspective, there are areas of naturalness which still await satisfactory implementation (Keller 2002). One area that has been identified is *expressive content*. When a human being speaks there is no fixed prosodic rendering for particular utterances. There are many ways of speaking the same sentence, and these are dependent on the various features of expression. It is important to stress that, whatever the source of expressive content in speech, it is an extremely changeable parameter. A speaker's expression varies within a few words, not just from complete utterance to complete utterance. With the present state of

Developments in Speech Synthesis   Mark Tatham and Katherine Morton
© 2005 John Wiley & Sons, Ltd.   ISBN: 0-470-85538-X

the art it is unlikely that a speech synthesiser will reflect *any* expression adequately, let alone one that is varying.

But to sound completely natural, speech synthesisers will sooner or later have to be able to reflect this most natural aspect of human speech in a way which convinces listeners that they could well be listening to real speech. This is one of the last frontiers of speech synthesis–and is so because it constitutes a near intractable problem.

There is no general agreement on what naturalness actually is, let alone on how to model it. But there are important leads in current research that are worth picking up and consolidating to see if we can come up with a way forward which will show promise of improved naturalness in the future. The work detailed in this book constitutes a *hypothesis*, a proposal for pushing speech synthesis forward on the naturalness front. It is not claimed in any sense that we are presenting the answer to the problem.

Many researchers agree that the major remaining obstacle to fully acceptable synthetic speech is that it continues to be insufficiently natural. Progress at the segmental level, which involves the perceptually acceptable rendering of individual segments and how they conjoin, has been very successful, but *prosody* is the focus of concern at the moment: the rendering of suprasegmental phenomena–elements that span multiple segments–is less than satisfactory and appears to be the primary source of perceptual unease. Prosody itself however is complex and might be thought of as characterising not just the basic prosody associated with rendering utterances for their plain meaning, but also the prosody associated with rendering the expressive content of speech. Prosody performs multiple functions– and it is this that needs particular attention at the moment. In this book one concern will be to address the issue of correct, or appropriate prosody in speech–not just the basic prosody but especially the prosody associated with expression.

Does synthetic speech improve on natural speech? According to some writers, for example Black (2002), there is a chance that some of the properties of speech synthesis can in fact be turned to advantage in some situations. For example, speech synthesisers can speak faster, if necessary, than human beings. This might be useful sometimes, though if the speech is faster than human speech it might be perceived or taken to be of lower quality. Philosophically this is an important point. We have now the means to convey information using something which is akin to human speech, but which could actually be considered to be an *improvement* on human speech. For the moment, though, this looks suspiciously like an explanation after the fact–turning a bug into a hidden feature! But this wouldn't be the first time in the history of human endeavour when we have had to admit that it possible to improve on what human beings are capable of.

Continuous Adaptation

The voice output of current synthesis systems does not automatically adapt to particular changes that occur during the course of a dialogue with a human being. For example, a synthetic utterance which begins with fast speech, ends with fast speech; and one which begins sounding firm does not move to a gentler style as the dialogue unfolds. Yet changes of this kind as a person speaks are a major property of naturalness in speech.

To simulate these changes for adequate synthesis we need a data structure characterisation sufficiently detailed to be able to handle dynamic changes of style or expression during the course of an utterance. We also need the means to introduce *marking* into the utterance specification which will reflect the style changes and provide the trigger for the appropriate procedures in the synthetic rendering.

The attributes of an utterance we are focussing on here are those which are rendered by the prosodic structure of the utterance. Prosody at its simplest implements the rhythm, stress and intonational patterns of canonical utterances. But in addition the parameters of prosody are used to render expressive content. These parameters are often characterised in a way which does not enable many of the subtleties of their use in human speech to be carried over to synthesis. For example, rate of delivery can vary considerably during the course of an utterance–a stretch of speech which might be characterised in linguistic terms as, say, a phrase or a sentence. Rate of delivery is a physical prosodic parameter which is used to render different styles that are characterised at an abstract level. For example, angry speech may be delivered at a higher than normal rate, bored speech at a lower than normal rate.

Take as an example the following utterance:

The word I actually used was *apostrophe*, though I admit it's a bit unusual.

In the orthographic representation of the word *apostrophe*, italicisation has been used to highlight it to indicate its infrequent use. In speech the word might

- be preceded and followed by a pause
- be spoken at a rate lower than the surrounding words
- have increased overall amplitude, and so on.

These attributes of the acoustic signal combine to throw spoken highlighting onto the word, a highlighting which says: *this is an unusual word you may not be familiar with*. In addition, uttering the word slowly will usually mean that phenomena associated with fast delivery (increased coarticulation, deliberate vowel reduction etc.) may not be present as expected. To a great extent it is the violation of the listener's expectations–dictated largely by the way the sentence has begun in terms of its prosodic delivery–which signals that they must increase their attention level here. What a speaker expects is itself a variable. By this we mean that there is a norm or baseline expectation for these parameters, and this in itself may be relative. The main point to emphasise is the idea of *departure from expectation*–whatever the nature or derivation of the expectation. In a sense the speaker *plays on* the listener's expectations, a concept which is a far cry from the usual way of thinking about speakers. We shall be returning frequently to the interplay between speaker and listener.

Data Structure Characterisation

Different synthesis systems handle both segmental and prosodic phenomena in different ways. We focus mainly on prosody here, but the same arguments hold for segmental phenomena. There is a good case for characterising the objects to be rendered in synthetic speech identically no matter what the special properties of any one synthesis system. Platform independence enables comparison and evaluation beyond the idiosyncrasies of each system.

Identical input enables comparison of the differing outputs, knowing that any differences detected have been introduced during the rendering process. For example:

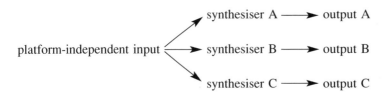

Note that outputs A, B and C can be compared with each other with respect to the common input they have rendered. Differences between A, B and C are therefore down to the individual characteristics of synthesisers A, B and C respectively.

The evaluation paradigm works only if the synthesisers under scrutiny are compliant with the characteristics of the input. Each may need to be fronted by a conversion process, and the introduction of this stage of processing is itself, of course, able to introduce errors in the output. But provided care is taken to ensure a minimum of error in the way the synthesiser systems enable the conversion the paradigm should be sound.

Applications in the field may need to access different synthesisers. In such a case a platform-independent high-level representation of utterances will go a long way to ensuring a minimum of disparity between outputs sourced from different systems. This kind of situation could easily occur, for example, with call centres switching through a hierarchy of options which may well involve recruiting subsystems that are physically distant from the initiating controller. The human enquirer will gain from the accruing continuity of output. However, common input, as we have seen, does not guarantee identity of output, but it does minimise discontinuities to which human users are sensitive. Indeed, since there are circumstances in which different synthesis systems are to be preferred–no single one is a universal winner–it helps a lot with comparison and evaluation if the material presented to all systems is identical.

Shared Input Properties

The properties in the input that are common are those which are quite independent of any subsequent rendering in the synthesis process. In general these properties are regarded as linguistic in nature, and in the linguistics model they precede phonetic rendering for the most part. By and large it is phonetic rendering which the low-level synthesis system is simulating. However, synthesis systems which incorporate high-level processing, such as text-to-speech systems, include phonological and other processing. Design of a platform-independent way of representing input to systems which incorporate some high-level processing is much more difficult than systems which involve only phonetic rendering. There are two possible polarised solutions, and several in between.

1 Remove from the text-to-speech system all processing more appropriately brought to a markup of the input text.
2 Introduce a platform-specific intermediate stage which removes only the least successful parts of the text-to-speech system and/or identifies any 'missing' processes.

The first solution implies standardising on high-level processes like text normalisation and orthography-to-phoneme conversion. This may not be a bad thing provided designers of text-to-speech systems were not compelled to drop their own processing in these areas if this were both compatible and at least as successful with respect to the final output. The problem really arises when we drop below this simple textual processing level and move into the linguistics processing proper–the phonology and in particular the prosody. We shall see later that whatever approach is adopted it becomes essential to identify what is to be drawn from an input markup and what is to be supplied in the text-to-speech system itself. The main way of avoiding confusion will be to have a common set of level identifiers to be used as markers indicating where in individual systems this or that process (to be included or rejected) occurs.

It will turn out to be important in the way we model human and synthetic speech production to distinguish between the linguistic properties of speech and the two main ways of rendering those properties: by human beings or by synthesisers. Within each of these two types there are different subtypes, and once again it is helpful if the input to all types is characterised in the same way. Along with linguistics in general, we claim a universality for the way in which all phenomena associated with speech production which *can* be characterised within linguistics (and perhaps some areas of psychology if we consider the perception of speech also) are to be described. What this means simply is that much is to be gained from adopting a universal framework for characterising important aspects of speech perception and production by both human beings and computers.

Intelligibility: Some Beliefs and Some Myths

A fairly common hypothesis among synthesis researchers is that the intelligibility of synthetic speech declines dramatically under conditions that are less than ideal. It is certainly true that when listening conditions are adverse synthetic speech appears to do less well than human speech as far as listeners are concerned–prompting the notion that human speech has more critical detail than synthetic speech. It follows from this that it can be hypothesised that adding the missing detail to synthetic speech will improve its intelligibility under adverse or more realistic listening conditions.

It is not self-evident, though, that increased detail is what is needed. For example it may well be that some systematic variation in human speech is not actually perceived (or used in the perception process) and/or it may be the case that some *non*-systematic detail is perceived, in the sense that if it is missing the result is the assertion that the speech is not natural. What constitutes naturalness is not entirely clear to anyone yet, if we go into this amount of detail in trying to understand what it means for speech to be intelligible to the point of being natural. Hence it becomes easy to equate naturalness with increased intelligibility and assign both to improved detail in the acoustic signal. If correct, then we go full circle on the observation that somehow or other human speech holds on to its intelligibility under adverse conditions where synthetic speech does not–even though both may be judged equally intelligible in the laboratory. Assertions of this kind are not helpful in telling us exactly what that detail of human speech might be; they simply inform us that human speech is perceptually more robust than synthetic speech, and that this is perhaps surprising if the starting point–the perception in the laboratory environment–is apparently equal. In our model (see Part V Chapter 20 and Part IX) we would hypothesise that the perceptual assignment process involves inputs that are not wholly those which on the face of it are responsible for intelligibility.

It is not difficult to imagine that formant synthesis may be producing a soundwave which is less than complete. The parameters for formant synthesis were selected in the early days–for example in the Holmes (1983) model–based on their obviousness in the acoustic signal and on their hypothesised relevance to perception. Thus parameters like formant peak frequency, formant amplitude and formant bandwidth were seen to be important, and were duly incorporated. Later systems–for example the Klatt (1980) synthesiser–built on this model to include parameters which would deliver more of the acoustic detail while attempting to maintain the versatility of formant or parametric synthesis. Fundamentally, it is quite true that, however carefully formant synthesis models the acoustic production of speech, the resultant signal is inevitably lacking in coherence and integrity. A speech signal with 100% integrity would require an infinite number of parameters to simulate it. The robust correlation between vocal tract behaviour and the detail of the acoustic signal is what makes natural speech acoustically coherent: its acoustic fine detail reflects vocal tract behaviour and identifies the signal as coming from a single talker. Indeed we could go further: the correlation is not just robust, it is probably *absolute*. What this correlation does not do on its own, however, is guarantee *phonetic* coherence, since vocal tract behaviour has a nonlinear relationship with phonetics and includes unpredictable cognitively sourced elements (Morton 1986; Tatham 1986a).

One or two researchers have taken a state-of-the-art parametric device–such as the Klatt synthesiser–and made the theoretical assumption that the coherence of its *output* can be improved by working on the internal integrity of its *input* (Stevens and Bickley 1991). HLSyn (Stevens 2002) is one such attempt. The proponents of HLSyn propose a level of representation which is intermediate between what is generally called high-level synthesis (corresponding to phonological, prosodic and pragmatic planning of utterances in linguistics) and low-level synthesis–the actual parametric device which creates the soundwave. They confuse the issue somewhat by calling HLSyn 'high-level synthesis', which is an idiosyncratic use of the term *high*. We shall see later that HLSyn and other comparable approaches (Werner and Haggard 1969; Tatham 1970a) do indeed introduce added coherence by linking acoustic detail *via* a shared higher level of representation–in this case an articulatory level (see also Mermelstein 1973). We would argue that for the moment it has not been shown that a similar level of coherence can be introduced by simply organising the acoustic parameters into an integrated structure.

Our own philosophy works roughly along these lines too: we are concerned with the integrity of the high-level parts of synthesis (rather than the intermediary levels which concern the HLSyn researchers). The principal example of this is our approach to prosody and expression–insisting that all utterance plans be wrapped in tightly focussed prosodic containers which ultimately control the rendering of temporal and spectral features of the output signal whether this is derived from a formant model or a concatenative waveform model.

But, certainly in our own experience, there are also similar though less severe problems with concatenated waveform synthesis, even in those systems which attempt to optimise unit length. This leads us to believe that although, of course, a certain minimum level of acoustic detail is necessary in all synthetic speech, the robustness issue is not down solely to a failure to replicate the greater spectral detail of human speech. What is left, of course, is prosodic detail and temporally governed variation of spectral detail. We are referring here to subtlety in fundamental frequency contours, and variations in intensity and rhythm for the prosodic detail *per se*; and also to the way spectral detail (for example, the variation in coarticulatory effects and the way they span much more than just the immediately adjacent segments) is governed by features like rate variation. These features are very complex when considering prosody in general, but particularly complex when considering prosody as conveyor of expression.

Naturalness

We shall be referring often in this book to natural sounding speech, and we share with many others an awareness of the vagueness of this idea. The perceived feeling of naturalness about speech is clearly based on a complex of features which it is difficult to enumerate. The reason for this is that listeners are unable to tell us precisely what contributes to naturalness. Several researchers have tried to introduce a metric for naturalness which goes beyond the simple marking of a scale, and introduces the notion of a parametric characterisation of what people feel as listeners. While not new of course in perceptual studies, such a method does go a long way toward enabling comparison between different systems by establishing the basis for a rough evaluation metric.

Take for example the naturalness scoring introduced by Sluijter *et al.* (1998). The approach is technically parametric and enumerates eleven parameters which listeners are asked to consider on five-point scales. They refer to these as a measure of acceptability, but acceptability and naturalness begin to converge in this type of approach–because of the idea that what is acceptable is also a prerequisite for what is natural. Sluijter *et al.*'s parameters can be readily glossed, adapted and extended:

1 *General quality*. What general impression does the speech create? In many studies this is the overall concept of naturalness and very often the only one evaluated.
2 *Ease of comprehension*. The question here for listeners is also general and elicits an overall impression of ease of comprehension. This parameter itself could be further para-meterised more objectively by a detailed analysis of what specifically causes problems of comprehension (as in the next feature, for example).
3 *Comprehension problems for individual words*. Here the listener can identify various difficult words, or in a more tightly controlled evaluation experiment the researchers can high-light words known to be difficult and try to analyse the reasons for the difficulties. For example, is there a semantic ambiguity with the word or is it phonologically similar to some other word and insufficiently disambiguated by the semantic context or the syntax?
4 *Intelligibility*. Once again, an overall ranking of general intelligibility.
5 *Pronunciation/occurrence of deviating speech sounds*. Does the listener feel that any particular sounds have been badly rendered and might be contributing to reduced naturalness or acceptability? Notice that errors in sounds in particular combinations will be less noticeable than in other combinations due to the predictability of linear sound combinations in syllables. How frequently do these rogue sounds occur?
6 *Speaking rate*. The question here is whether the speaking rate is appropriate. One difficulty is the extent to which semantic and pragmatic factors enter into the appro-priateness of speaking rate. Most synthesisers have a default speaking rate, and maybe should be evaluated on this. Introducing variation of speaking rate may well introduce errors. This is one of the areas–along with other pragmatically sourced variations in

prosody–which will benefit from additional markup of the input text or superior prosody assignment algorithms within the synthesis system.

7 *Voice pleasantness*. A general and very impressionistic parameter, and one which might vary with semantic and prosodic content.

8 *Naturalness*. A general parameter which it may be possible to refine a little. So, we may be able to ask questions like: Is the acoustics of the utterance internally coherent? For example:

- *Does the speech appear to be from a single speaker?*
- *Does this coherence extend throughout the fundamental frequency range with an appropriate amplitude dynamics?*

9 *Liveliness*. In general, liveliness is judged to be a desirable quality contributing to naturalness. But it could be argued that for a general system a whole range of expression along a dullness–liveliness vector should be possible, derived either internally or in response to markup. So the question here is really not

- *Is the speech lively?* but rather
- *Is the degree of liveliness applied to an appropriate degree?*

10 *Friendliness*. This is a quality appropriate for limited domain systems–say, interactive enquiry systems. But in a general system it would be subject, as with naturalness, liveliness and politeness (below), to semantic and pragmatic content. Appropriateness is again a consideration after determining that the default degree of friendliness is convincing.

11 *Politeness*. Again, a subjective evaluation of the default condition–degree of politeness in general–is called for. But also appropriateness for content, and a suitable interpretation of markup, if present, are required judgements.

Each of these parameters is subjective and again defined only vaguely for that reason; and not enough provision is made for adaptation on the part of the listener. But the strength of such an approach is that, notwithstanding the subjective nature of each parameter, the evaluation of naturalness *as a whole* is made more robust. This stems in part from modelling in terms of identifiable features to which a probability might be attached, and in part from the possibility of indicating a relationship between the features. Whilst far from robust in a fully objective way, the characterisation of naturalness here does gain over a non-parametric characterisation, and the approach may eventually lead to productive correlation between measured properties of the soundwave and naturalness. The effects of rendering markup would be an appropriate application for this evaluation technique.

Systems are now good enough for casual listeners to comment not so much on naturalness but on the *appropriateness* of style–as with the friendliness and politeness parameters used by Sluijter *et al*. This does not mean that style, and allied effects, are secondary to naturalness in terms of generating speech synthesis, but it does mean that for some people appropriateness and accuracy of style override some other aspects of naturalness. These considerations would not override intelligibility, which still stands as a prerequisite.

Variability

One of the paradoxes of speech technology is the way in which variability in the speech waveform causes so many problems in the design of automatic speech recognition systems and at the same time *lack* of it causes a feeling of unnaturalness in synthesised speech. Synthesis seeks to introduce the variability which recognition tries to discard.

Linguistics models variability in terms of a hierarchical arrangement of identifiably different types. We discuss this more fully in Chapter 14, but for the moment we can recognise:

- deliberately introduced and systematic–*phonology*
- unavoidable, but systematic (coarticulation)–*phonetics*
- systematically controlled coarticulation–*cognitive phonetics*
- random–*phonetics*

1 The variability introduced at the phonological level in speech production involves the introduction by the speaker of variants on the underlying segments or prosodic contours. So, for example, English chooses to have two non-distinctive variants of /l/ which can be heard in words like *leaf* and *feel*–classical phonetics called these clear [l] and dark [l] respectively. In the prosody of English we could cite the variant turning-up of the intonation contour before the end of statements as opposed to the usual turn-down. Neither of these variants alters the basic meaning of the utterance, though they can alter pragmatic interpretation. These are termed *extrinsic* variants, and in the segment domain are called *extrinsic allophones*. Failure to reproduce phonological variability correctly in synthetic speech results in a 'foreign accent' effect because different languages derive extrinsic allophones differently; the meaning of the utterance however is not changed, and it usually remains intelligible.

2 Segmental variants introduced unavoidably at the phonetic level are termed *intrinsic allophones* in most contemporary models of phonetics and result from coarticulation. Coarticulation is modelled as the distortion of the intended articulatory configuration associated with a segment–its target–by mechanical or aerodynamic inertial factors which are intrinsic to the speech mechanism and have nothing to do with the linguistics of the language. These inertial effects are systematic and time-governed, and are predictable. Examples from English might be the fronted [k] in a word like *key*, or the dentalised [t] in *eighth*; or vocal cord vibration might get interrupted during intervocalic underlying [+voice] stops or fricatives. Failure to replicate coarticulation correctly in speech synthesis reduces overall intelligibility and contributes very much to lack of naturalness. Interestingly, listeners are not aware of coarticulatory effects in the sense that they cannot report them: they are however extremely sensitive to their omission and to any errors.

3 Observations of coarticulation reveal that it sometimes looks as though coarticulatory effects do vary in a way related to the linguistics of the language, however. The most appropriate model here for our purposes borrows the notion of *cognitive intervention* from bio-psychology to introduce the idea that within certain limits the mechanical constraints can be interfered with–though rarely, if ever, negated completely. Moreover it looks as though some effects intrinsic to the mechanism can actually be enhanced at will for linguistic purposes. Systematic cognitive intervention in the behaviour of the physical mechanism which produces the soundwave is covered by the theory of cognitive phonetics (see Chapter 27). Examples here might be the way coarticulation is reduced in any language when there is a high risk of ambiguity–the speaker slows down to reduce the time-governed constraint–or the enhanced period of vocal cord vibration

failure following some stops in a number of Indian languages. This cognitive intervention to control mechanical constraints enables the enlargement of either the language's extrinsic allophone inventory or even sometimes its underlying segment (phoneme) inventory. If the effects of cognitive intervention in phonetic rendering are not reproduced in synthetic speech there can be perceptual problems occasionally with meaning, and frequently with the coherence of accents within a language. There is also a fair reduction in naturalness.

4 Some random variability is also present in speech articulation. This is due to tolerances in the mechanical and aerodynamic systems: they are insufficiently tight to produce error- or variant-free rendering of the underlying segments (the extrinsic allophones) appearing in the utterance plan. While listeners are not at all sensitive to the detail of random variability in speech, they do become uneasy if this type of variability is not present; so failure to introduce it results in a reduction of naturalness.

Most speech synthesis systems produce results which take into account these types of variability. They do, however, adopt widely differing theoretical stances in how they introduce them. In stand-alone systems this may not matter unless it introduces errors which need not otherwise be there. However, if we attempt to introduce some cross-platform elements to our general synthesis strategy the disparate theoretical foundations may become a problem. In Chapter 20 and Chapter 32 we discuss the introduction of prosodic markup of text input to different synthesis systems. There is potential here for introducing concepts in the markup which may not have been adopted by all the systems it is meant to apply to. A serious cost would be involved if there had to be alternative front ends to copy for different theoretical assumptions in the markup.

Variability is still a major problem in speech synthesis. Linguists are not entirely in agreement as to how to model it, and it may well be that the recognition of the four different types mentioned above rests on too simplistic an approach. Some researchers have claimed that random variability and cognitively controlled intrinsic effects are sometimes avoided or minimised in order to improve intelligibility; this claim is probably false. Cognitive intervention definitely contributes to intelligibility and random variation definitely contributes to naturalness; and intelligibility and naturalness are not entirely decoupled parameters in the perception of speech. It is more likely that some areas of variability are avoided in some synthesis because of a lack of data. In successful state-of-the-art systems, variability is explicitly modelled and introduced in the right places in the planning and rendering algorithms.

The Introduction of Style

Although the quality of text-to-speech systems is improving quite considerably, probably due to the widespread adoption of concatenative or unit selection systems, most of these systems can speak only with one particular style and usually only one particular voice. The usual style adopted, because it is considered to be the most general-purpose, is a relatively neutral version of reading-style speech. What most researchers would like to see is the easy extension of systems to include a range of voices, and also to enable various global styles and local expressive content. All these things are possible–but not yet adopted in systems outside the laboratory.

Prosody control is essential for achieving different styles within the same system. Speech rate control is a good place to start for most researchers because it gives the appearance of being easy. The next parameter to look at might be fundamental frequency or intonation. However, the introduction of speech rate control turns out to be far from simple. The difficulty is expressed clearly in the discovery that a doubling of overall rate is not a halving of the time spent on each segment in an utterance–the distribution of the rate increase is not linear throughout the utterance. Focussing on the syllable as our unit we can see that a change in rate is more likely to affect the vowel nucleus than the surrounding consonants, but it is still hard to be consistent in predicting just how the relative distribution of rate change takes effect. We also observe (Tatham and Morton 2002) that global rate change is not reflected linearly in the next unit up either–the rhythmic unit. A rhythmic unit must have one stressed syllable which begins the unit. Unstressed syllables between stressed ones are fitted into the rhythmic unit, thus:

<utterance>| Pro.so.dy.is | prov.ing | hard.to | mo.del | ac.cur.ate.ly | </utterance>

Here rhythmic unit boundaries are marked with 'l' and stressed syllables are underlined. A '.' separates syllables.

The usual model acknowledges the perceptually oriented idea of isochrony between rhythmic units, though despite proving a useful concept in phonological prosody (that is, in the abstract) it is hard to find direct correlates in phonetic prosody–that is, in the actual soundwave. The isochrony approach would hypothesise that the perceived equal timing between stressed syllables–the time between the vertical markers in the above representation–is reflected in the physical signal. The hypothesis has been consistently refuted by researchers.

Evaluating the segmental intelligibility of synthesisers neglects one feature of speech which is universally present–expressive content. In the early days of synthesis the inclusion of anything approaching expression was an unaffordable luxury–it was difficult enough to make the systems segmentally intelligible. Segmental intelligibility, however, is no longer an issue. This means that attention can be directed to evaluating expressive content. In the course of this book we shall return many times to the discussion of expression in synthesis, beginning with examining just what expression *is* in speech. But even if our understanding of expression were complete it would still be difficult to test the intelligibility of synthesised expression. We do not mean here answering questions like 'Does the synthesiser sound happy or angry?' but something of a much more subtle nature. Psychologists have researched the perception of human sourced expression and emotion, but testing and evaluating the success of synthesising expressiveness is something which will have to be left for the future for the moment.

Expressive Content

Most researchers in the area of speech synthesis would agree that the field has its fair share of problems. What we decided when planning this book was that for us there are three major problems which currently stand out as meriting research investment if real headway is to be made in the field *as a whole*. All researchers will have their own areas of interest, but these are our personal choice for attention at the moment. Our feeling is that these three areas contribute significantly to whether or not synthetic speech is judged to be *natural*–and for us it is an overall improvement in naturalness which will have the greatest returns in the near future. Naturalness therefore forms our *first problem area.*

Naturalness for us hinges on expressive content–expression or the lack of it is what we feel does most to distinguish current speech synthesis systems from natural speech. We shall discuss later

- whether this means that the acoustic signal must more accurately reflect the speaker's expression, or
- whether the acoustic signal must provide the listener with cues for an accurate perceiver assignment of the speaker's *intended* expression.

We shall try to show that these are two quite different things, and that neither is to be neglected. But we shall also show that most current systems are not well set up for handling expressive content. In particular they are not usually able to handle expression on a dynamic basis. We explain why there is a need for dynamic modelling of expression in speech synthesis systems. Dynamic modelling is our *second problem area.*

But the approach would be lacking if we did not at the same time show a way of integrating the disparate parts of a speech synthesis system which have to come together to achieve these goals. And this is our *third problem area*–the transparent integration of levels within synthesis.

The book discusses current work on high-level synthesis, and presents proposals for a unified approach to addressing formal descriptions of high-level manipulation of the low-level synthesis systems, using an XML-based formalism to characterise examples. We feel XML is ideally suited to handling the necessary data structures for synthesis. There were a number of reasons for adopting XML; but mostly we feel that it is an appropriate markup system for

- characterising data structures
- application on multiple platforms.

One of the important things to realise is that modelling speech production in human beings or simulating human speech production using speech synthesis is fundamentally a problem of characterising the data structures involved. There are procedures to be applied to these data structures, of course; but there is much to be gained from making the data structures themselves the focus of the model, making procedures *adjunct* to the focus. This is an approach often adopted for linguistics, and one which we ourselves have used in the SPRUCE model and elsewhere (Tatham and Morton 2003, 2004) with some success.

The multiple-platform issue is not just that XML is interpretable across multiple operating systems or 'in line' within different programming languages, but more importantly that it can be used to manage the high-level aspects of speech synthesis in text-to-speech and other speech synthesis systems which hinge on high-level aspects of synthesis. Thus a high-level approach can be built which can precede multiple low-level systems. It is in this particular sense that we are concerned with the application across multiple platforms. As an example we can cite the SPRUCE system which is essentially a high-level synthesis system whose output is capable of being rendered on multiple low-level systems–such as formant-based synthesisers or those based on concatenated waveforms.

Final Introductory Remarks

The feeling we shall be trying to create in this book is one of optimism. Gradually inroads are being made in the fields of speech synthesis (and the allied field of automatic speech recognition) which are leading to a greater understanding of just how complex human speech and its perception are. The focus of researchers' efforts is shifting toward what we might call the *humanness* of human speech–toward gaining insights into not so much how the message is encoded using a small set of sounds (an abstract idea), but how the message is cloaked in a multi-layered wrapper involving continuous adjustment of detail to satisfy a finely balanced interplay between speaker and listener. This interplay is most apparent in dialogue, where on occasion its importance exceeds even that of any messages exchanged. Computer-based dialogue will undoubtedly come to play a significant role in our lives in the not too distant future. If conversations with computers are to be at all successful we will need to look much more closely at those areas of speaker/listener interaction which are beginning to emerge as the new focal points for speech technology research.

We have divided the book into a number of parts concentrating on different aspects of speech synthesis. There are a number of recurrent themes: sometimes these occur briefly, sometimes in detail–but each time from a different perspective. The idea here is to try to present an integrated view, but from different areas of importance. We make a number of proposals about approach, modelling, the characterisation of data structures, and one or two other areas: but these have only the status of suggestions for future work. Part of our task has been to try to draw out an understanding of why it is taking so long to achieve genuinely usable synthetic speech, and to offer our views on how work might proceed.

So we start in Part I with establishing a firm distinction between high- and low-level synthesis, moving toward characterising naturalness as a new focus for synthesis in Part II. We make suggestions for handling high-level control in Part III, highlighting areas for improvement in Part IV. Much research has been devoted recently to markup of text as a way of improving detail in synthetic speech: we concentrate in Part V on highlighting the main important advances, indicating in Part VI some detail of how data structures might be handled from both static and dynamic perspectives. A key to naturalness lies in good handling of prosody, and in Part VIII we move on to some of the details involving in coding and rendering, particularly of intonation. We present simple ways of handling data characterisation in XML markup, and its subsequent processing with examples in procedural pseudo-code designed to suggest how the various strands of information which wrap the final signal might come together. Part IX pulls the discussion together, and the book ends with a concluding overview where we highlight aspects of speech synthesis for development and improvement.

Part I

Current Work

1

High-Level and Low-Level Synthesis

1.1 Differentiating Between Low-Level and High-Level Synthesis

We need to differentiate between low- and high-level synthesis. Linguistics makes a broadly equivalent distinction in terms of human speech production: low-level synthesis corresponds roughly to phonetics and high-level synthesis corresponds roughly to phonology. There is some blurring between these two components, and we shall discuss the importance of this in due course (see Chapter 11 and Chapter 25).

In linguistics, phonology is essentially about planning. It is here that the plan is put together for speaking utterances formulated by other parts of the linguistics, like semantics and syntax. These other areas are not concerned with speaking–their domain is how to arrive at phrases and sentences which appropriately reflect the meaning of what a person has to say. It is felt that it is only when these pre-speech phrases and sentences are arrived at that the business of planning how to speak them begins. One reason why we feel that there is somewhat of a break between sentences and planning how to speak them is that those same sentences might be channelled into a part of linguistics which would plan how to *write* them. Diagrammed this looks like:

The task of phonology is to formulate a speaking plan, whereas that of graphology is to formulate a writing plan. Notice that in either case we end up simply with a plan–not with writing or speech: that comes later.

1.2 Two Types of Text

It is important to note the parallel between graphology and phonology that we can see in the above diagram. The apparent equality between these two components conceals something very important, which is that the graphology pathway, leading eventually to a rendering of the sentence as writing, does not encode as much of the information available

Developments in Speech Synthesis   Mark Tatham and Katherine Morton
© 2005 John Wiley & Sons, Ltd.   ISBN: 0-470-85538-X

at the sentence level as the phonology does. Phonology, for example, encodes prosodic information, and it is this information in particular which graphology barely touches.

Let us put this another way. In the human system graphology and phonology assume that the recipient of their processes will be a human being–it has been this way since human beings have written and spoken. Speech and writing are the result of systematic renderings of sentences, and they are intended to be decoded by human beings. As such the processes of graphology and phonology (and their subsequent low-level rendering stages: 'graphetics' and phonetics) make assumptions about the device (the human being) which is to input them for decoding.

With speech synthesis (designed to simulate phonology and phonetic rendering), text-to-speech synthesis (designed to simulate human beings reading aloud text produced by graphology/graphetics) and automatic speech recognition (designed to simulate human perception of speech) such assumptions *cannot* be made. There is a simple reason for this: we really do not yet have adequate models of all the human processes involved to produce other than imperfect simulations.

Text in particular is very short on what it encodes, and as we have said, the shortcomings lie in that part of a sentence which would be encoded by prosodic processing were the sentence to be spoken by a human being. Text makes one of two assumptions:

- the text is not intended to be spoken, in which case any expressive content has to be text-based–that is, it must be expressed using the available words and their syntactic arrangement;
- the text is to be read out aloud; in which case it is assumed that the reader is able to supply an appropriate expression and prosody and bring this to the speech rendering process.

By and large, practised human readers are quite good at actively adding expressive content and general prosody to a text while they are actually reading it aloud. Occasionally mistakes are made, but these are surprisingly rare, given the look-ahead strategy that readers deploy. Because the process is an active one and depends on the speaker and the immediate environment, it is not surprising that different renderings arise on different occasions, even when the same speaker is involved.

1.3 The *Context* of High-Level Synthesis

Rendering processes within the overall text-to-speech system are carried out within a particular context–the prosodic context of the utterance. Whether a text-to-speech system is trying to read out text which was never intended for the purpose, or whether the text has been written with human speech rendering in mind, the task for a text-to-speech system is daunting. This is largely down to the fact that we really do not have an adequate model of what it is that human readers bring to the task or how they do it. There is an important point that we are developing throughout this book, and that is that it is not a question of *adding* prosody or expression, but a question of rendering a spoken version of the text *within* a prosodic or expressive framework. Let us term these the *additive model* and the *wrapper model*. We suggest that high-level synthesis–the development of an utterance plan–is conducted within the wrapper context.

Conceptually these are very different approaches, and we feel that one of the problems encountered so far is that attempts to *add* prosody have failed because the model is too simplistic and insufficiently reflective of what the human strategy is. This seems to focus on rendering speech within an appropriate prosodic wrapper, and our proposals for modelling the situation assume a hierarchical structure which is dominated by this wrapper (see Chapter 34).

Prosody is a general term, and can be viewed as extending to an abstract characterisation of a vehicle for features such as expressive or, more extremely, emotive content. An abstract characterisation of this kind would enumerate all the possibilities for prosody, and part of the rendering task would be to highlight appropriate possibilities for particular aspects of expression. Notice that it would be absurd to attempt to render 'prosody' in this model (since it is simultaneously *everything* of a prosodic nature), just as it is absurd to try to render syntax in linguistic theory (since it simultaneously characterises *all possible* sentences in a language). Unfortunately some text-to-speech systems have taken this abstract characterisation of prosody and, calling it 'neutral' prosody, have attempted to add it to the segmental characterisation of particular utterances. The results are not satisfactory because human listeners do not know what to make of such a waveform: they know it *cannot* occur. Let us summarise what is in effect a matter of principle in the development of a model within linguistic theory:

- Linguistic theory is about the knowledge of language and of a particular language which is shared between speakers and listeners of that language.
- The model is static and does not include–in its strictest form–processes involving drawing on this knowledge for characterising particular sentences.

As we move through the grammar toward speech we find the linguistic component referred to as *phonology*–a characterisation for a particular language of everything necessary to build utterance plans to correspond with the sentences the grammar enumerates (all of them). Again there is no formal means within this type of theory for drawing on that knowledge for planning specific utterances.

Within the phonology there is a characterisation of prosody, a recognisable sub-component–intonational, rhythmic and prominence features of utterance planning. Again prosody enumerates all possibilities and, we say again, with no recipe for drawing on this knowledge. Although prosody is normally referred to as a sub-component of phonology, we prefer to regard phonological processes as taking place within a prosodic context: that is, prosodic processes are logically prior to phonological processes. Hence the wrapper model referred to above.

Pragmatics is a component of the theory which characterises expressive content–along the same theoretical lines as the other components. It is the interaction between pragmatics and prosody which highlights those elements of prosody which associate with particular pragmatic concepts. So, for example, the pragmatic concept of anger (deriving from the bio-psychological concept of anger) is associated with features of prosody which when combined uniquely characterise expressive anger. Prosody is not, therefore, 'neutral' expression; it the characterisation of all possible prosodic features in the language.

1.4 Textual Rendering

Text-to-speech systems, by definition, take written text which would have been derived from a writing plan devised by a 'graphology', and use it to generate a speaking plan which is then spoken. Notice from the diagram above, though, that human beings obviously do not do this; writing is an *alternative* to speaking, it does not precede it. The exception, of course, is when human beings themselves take text and read it out loud. And it is this human behaviour which text-to-speech systems are simulating.

We shall see that an interesting problem arises here. The operation of graphology–the production of a plan for writing out phrases or sentences–constitutes a 'lossy' encoding process: information is lost during the process. What eventually appears on paper does not encode all of a speaker's intentions. For example, the mood of the writer does not come across, except perhaps in the choice of particular words. The mood of a *speaker*, however, invariably *does* come across. Immediately, though, we can observe that mood (along with emotion or intention) could not have been encoded in the phrase or sentence except *via* actual words–so much of what a third party detects of a person's mood is conveyed by tone-of-voice. It is not actually expressed in the sentence to begin with.

The human system gets away with this lossy encoding that we come across in written text because human readers in general find no difficulty in restoring what has been removed–or at least some acceptable substitute. For example, compare the following two written sentences:

It was John.
It wasn't Mary, it was John.

The way in which the words *It was John* are spoken differs, although the text remains the same. No native speaker of English who can also read fails to make this difference. But a text-to-speech system struggles to add the contrastive emphasis which a listener would expect. This is an easy example–it is not hard to imagine variations in rendering an utterance which are much more subtle than this.

Some researchers have tried to estimate what is needed to perform this task of restoring semantic or pragmatic information at the reading aloud stage. And to a certain extent restoration is possible. But most agree that there are subtleties which currently defeat the most sophisticated algorithms because they rest on unknown factors such as *world knowledge*– what a speaker knows about the world way beyond the current linguistic context.

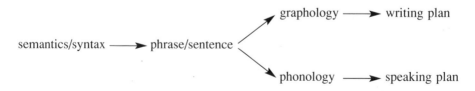

The above diagram is therefore too simple. There is a component missing–something which accounts for what a speaker, in planning and rendering an utterance, brings to the process which would not normally be encoded. A more appropriate diagram would look like this:

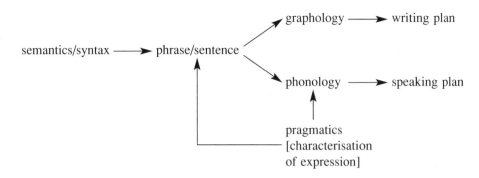

In linguistics, much of a person's expression (mood, emotion, attitude, intention) is characterised by a component called *pragmatics*, and it is here that the part of language which has little to do with choice of actual words is formulated. Pragmatics has a direct input into phonology (and phonetics) and influences the *way* in which an utterance is actually spoken.

Pragmatics (Verschueren 2003) is maturing late. The semantic and syntactic areas matured earlier, as well as phonology (without reference to expression–the phonology of what has been called the *neutral* or expressionless utterance plan). It was therefore not surprising that the earlier speech technology models, adopted in speech synthesis and automatic speech recognition, were not sensitive to expression–they omitted reference to pragmatics or its output. We reiterate many times in this book that a major reason for the persistent lack of convincing naturalness in speech synthesis is that systems are based on a pragmatics-free model of linguistics.

> A pragmatics-free model of linguistics fails to accommodate the variability associated with what we might call in very general terms expression or tone-of-voice. The kinds of things reflected here are a speaker's emotional state, their feelings toward the person they're speaking to, their general attitude, the environmental constraints which contribute to the choice of style for the speech, etc. There are many facets to this particular problem which currently preoccupy many researchers (Tatham and Morton 2004).

One of the tasks of this book will be to introduce the theoretical concepts necessary to enable an expression information channel to link up with speech planning, and to show how this translates into a better plan for rendering into an actual speech soundwave. Although this book is not about automatic speech recognition, we suggest that consideration of expression, properly modelled in phonology and phonetics, points to a considerable improvement in the performance of automatic speech recognition systems.

2

Low-Level Synthesisers: Current Status

2.1 The Range of Low-Level Synthesisers Available

Although the main thrust of this book concerns high-level synthesis, we will enumerate the most common types of low-level synthesiser, showing their main strengths and weaknesses in terms of the kind of information they need to handle to achieve greater naturalness in their output. This is not a trivial matter: the synthesiser types are not equivalent, and some are better than others in rendering important properties of human speech.

There are main three categories of speech synthesis technique:

- articulatory synthesis
- formant synthesis
- concatenative synthesis.

2.1.1 Articulatory Synthesis

Articulatory synthesis aims to simulate computationally the neurophysiology and biomechanics of speech production. This is perhaps the least developed technique, though potentially it promises to be the most productive in terms of accuracy, and therefore in terms of recreating the most detail in the soundwave. Articulatory synthesis has been heralded as the holy grail of synthesis precisely because it would simulate all aspects of human speech production below the level of the utterance plan. We are confident that phoneticians do not yet have a speech production model which can support full articulatory synthesis, but a comprehensive system would need to include several essential components, among them:

- *An input* corresponding, in physical terms, to the utterance plan generated cognitively in a speaker. The representation here would need to be compatible both with high-level cognitively based processes and lower level neurophysiological and biomechanical processes.
- *An adequate model of speech motor control*, which may or may not differ from other forms of motor control (such as walking) in the human being. Motor control involves not only the direct signalling of motor goals to the musculature of articulation and pulmonary control, but mechanisms and processes covering reflex and cognitively processed feedback

Developments in Speech Synthesis   Mark Tatham and Katherine Morton
© 2005 John Wiley & Sons, Ltd.   ISBN: 0-470-85538-X

(including proprioceptive and auditory feedback). Intermuscular communication would need to be covered and its role in speech production explained.

- A *biomechanical model* able to predict response and inertial factors to explain staggered enervation of high-mass objects and low-mass objects to achieve simultaneity of articulatory parameters contributing to the overall gestural goal.
- A *model of the aerodynamic and acoustic properties of speech*, especially of laryngeal control and function. The model would need to explain both source and filter characteristics of speech production, including differences between voices. Different types of phonation would need to be covered.

The task facing the developers of a low-level articulatory synthesiser are daunting, though there have been one or two research systems showing encouraging success (Rubin *et al.* 1981; Browman and Goldstein 1986). A later attempt to include some articulatory properties in speech synthesis comes from the HLSyn model (Stevens 2002). HLSyn stands for *high-level synthesis*–though paradoxically the model is concerned only with *low-level* (i.e. physical) processes, albeit of an articulatory rather than acoustic nature. The developers of HLSyn seem to have been driven less by the desire to simulate articulation in human beings than by the desire to provide an abstract level just above the acoustic level which might be better able to bring together features of the acoustic signal to provide synthetic speech exhibiting greater than usual integrity. The advocates of introducing what is in fact a representational articulatory layer within the usual cognitive-to-acoustic text-to-speech system claim that the approach does indeed improve naturalness, primarily by providing a superior approach to coarticulation. The approach begs a major question, however, by assuming that the best model of speech production involves coarticulated concatenative objects–a popular approach, but certainly not the only one; see Browman and Goldstein (1986) and the discussion in Part IV Chapter 2).

2.1.2 Formant Synthesis

A focal point for differentiating different low-level synthesisers is to look at the way in which data for the final waveform is actually stored. This ranges from multiple databases in some concatenated waveform systems to cover multi-voices and different modes of expression, to the minimalist segment representational system of the Holmes and other early formant synthesis systems.

The Holmes text-to-speech system (Holmes *et al.* 1964)–and those like it, for example MITalk (Allen *et al.* 1987)–is interesting because of the way it encapsulates a particular theory of speech production. In this system representations of individual speech segments are stored on a parametric basis. The parameters are those of the low-level Holmes formant synthesiser (Holmes 1983). For each segment there is a single value for each parameter. This means that the representation is abstract, since even if it was possible to represent a single acoustic segment of speech we would not choose a single representation because segments would (even in isolation) *unfold* or develop in a way which would probably alter the representation as time proceeded through the segment. If we look at the accompanying spectrogram (Figure 2.1) of a single segment which can in natural speech occur on its own we find that the signal is by no means 'static'–there is a clear dynamic element. This is what we meant by saying that the segment unfolds.

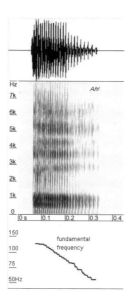

Figure 2.1 Waveform and spectrogram of the word *Ah!*, spoken in isolation and with falling intonation. Note that as the utterance unfolds there are changes in its spectral content and overall amplitude. These parameters would normally be held constant in a text-to-speech system using formant synthesis. All waveform, spectrogram and fundamental frequency displays in this book were derived using Mark Huckvale's Speech Filing System (www.ucl.ac.uk/phonetics).

The Holmes representation is abstract because it does not attempt to capture the entire segment; instead, it attempts to represent the segment's *target*–supposedly what the speaker is aiming for in this particular theory of speech production: *coarticulation theory* (Hardcastle and Hewlett 1999). Variations or departures from this target during the actual resulting waveform are (often tacitly) thought of as departures due to imperfections in the system. One way of quantifying what is in fact an abstract concept is to think of the target as being the average parameter values for the entire segment. So, we might represent [ɑ] as shown in Table 2.1.

The final average column in the table is calculated from several measured values sampled across the example segment. For each of the segments needing target representations in the language, a set of target parameter values is provided. A final parameter in the target representation would be the average duration of this particular segment in the language. Averages of data are *not* the data themselves–they result from a data reduction process and are therefore abstractions from the data. So these target representations are abstract and never actually occur. If any acoustic signal really does have these values then it is simply just another value. In target theory we could say that the best way of thinking of an actual signal is to say that it has been derived from the abstract target representation, and that a focus of interest would be the different types of derivation and their actual detail. Most of the derivation processes would be systematic, but in addition (almost a *sine qua non* since we are dealing with human beings) there is a certain random element. A model of this kind would account for the way in which the signal of a single speech sound varies throughout its lifetime and accounts for the way in which we characterise that signal–as a sequence of derivations based on a single abstract target.

Table 2.1

Sample	fn	alf	F1	a1	F2	a2	F3	a3	ahf	s	f0
1	250	61	725	63	1000	63	2450	54	55	63	26
2	250	60	725	63	1000	63	2450	61	60	63	26
3	250	56	700	60	950	61	2900	56	62	63	27
4	250	55	725	59	950	63	2800	54	60	63	26
5	250	55	750	59	900	63	2900	53	58	63	25
6	250	55	725	59	950	63	2850	46	57	63	24
7	250	54	700	58	900	58	2900	47	53	63	23
8	250	53	700	57	950	59	2900	49	48	45	24
9	250	52	725	56	950	63	2900	48	49	45	23
10	250	51	700	55	900	62	2900	47	50	63	20
11	250	49	700	54	950	61	2800	46	48	63	20
12	250	48	750	53	950	59	2600	45	55	45	21
13	250	48	725	51	1000	57	2850	39	55	63	22
14	250	36	750	37	1050	48	2900	48	50	63	25
15	250	30	775	29	1025	49	2900	45	49	45	23
Mean	**250**	**51**	**725**	**54**	**963**	**57**	**2800**	**49**	**54**	**61**	**24**
Target	**250**	**51**	**725**	**54**	**975**	**57**	**2800**	**49**	**54**	**63**	**24**

The table shows 15 samples taken at 10-ms intervals during an isolated utterance [ɑ]. The eleven parameters are: **fn**, a parameter used for handling nasal sounds; **alf**, amplitude of low frequencies (works in conjunction with fn); **F1**, frequency of F1; **a1**, amplitude of F1; **F2**, frequency of F2; **a2**, amplitude of F2; **F3**, frequency of F3; **a3**, amplitude of F3; **ahf**, amplitude of high frequencies (works to provide high formant amplitudes and the amplitudes of some fricatives); **s**, a source type switch; **f0**, the fundamental frequency.

alf, *a1*, *a2*, *a3* and *ahf* are expressed in ranges of 64 levels. For *s*, 63 = periodic source, 45 = mixed source, and 1 = aperiodic source. *f0* has 64 levels with 1 = 27Hz and 63 = 400Hz. Each time frame is 10ms. Details are in Holmes *et al.* (1964).

Notice that although all values are derived from just 150ms of the vowel they do show some variation as the vowel unfolds. This is taken care of by averaging the values to provide arithmetic means which constitute the abstract target. If the isolated vowel is now synthesised in a typical text-to-speech system, the final waveform would be based on 15 repetitions of the target, with the original variability removed. Listeners are sensitive to lack of variability where they expect it and the output would immediately sound unnatural.

However, a repetition of the same sounds or words produces different signals. The accompanying spectrograms (Figure 2.2) are of the same speaker speaking the utterance *Notice how much variability there is* three times. We now have variation of the target within each repetition and between repetitions. And once again the processes involved in accounting for these variations will be in part systematic and predictable and in part random. Having different speakers repeat the same sound will widen the variability again.

This, of course, is the well-known variability problem of automatic speech recognition. Many researchers in the automatic speech recognition field still feel their major task to be the *reduction* of these variations down to a single, identifiable abstract target corresponding to what the model feels to be the speaker's *intended* sound. Very recently

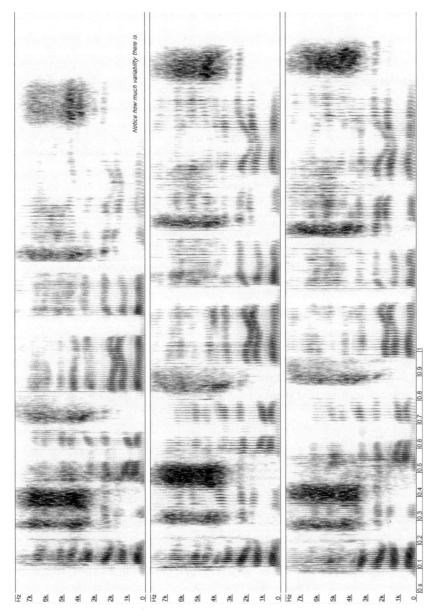

Figure 2.2 Spectrograms of the utterance *Notice how much variability there is*, spoken three times by the same speaker. Overall utterance duration varies, but individual elements within the utterance vary their durations differently in the three versions. Formant bending associated with coarticulatory processes also varies for the same elements in the different versions.

researchers in speech synthesis have turned to ways of reproducing this variability in the signal output from the synthesiser. The reason for doing this is that listeners to synthetic speech are very sensitive to failure to include such variability. One reason why early synthesisers sounded so unnatural was that any particular utterance would be repeated with a near identical waveform–something listeners do not expect from human speakers (see the spectrograms).

2.1.3 Concatenative Synthesis

Units for Concatenative Synthesis

We begin the discussion on low-level concatenative synthesis by considering the earliest forms: diphone synthesis and synthesis based on similar units. The unifying link here is the theoretical basis for the units: the modelling of speech as a sequence of otherwise individual sound segments or phones. In diphone-based systems a database of all possible diphones in the language is pre-stored and novel output is obtained by recombining appropriate diphones called from the database. However, various early systems also used half-syllables, triphones and phones as their building blocks. Figure 2.3 illustrates these units as they might be marked for excision from a sample of continuous speech.

- A *phone* is a single unit of sound. The idea that speech is a sequence of such sounds, albeit overlapping or blended together by coarticulatory processes, derives in modern times from classical phonetics. The principle of the isolable sound is a cornerstone of this theory.
- A *diphone* is defined as the signal from either the mid point of a phone or the point of least change within the phone to the similar point in the next phone.
- A *half-syllable* is like a diphone, except that the unit is taken from the mid point or point of least change in the phonetic syllable to a similar point in the next syllable.
- A *triphone* is a section of the signal taking in a sequence going from the middle of a phone, completely though the next one to the middle of a third. As with diphones, the segmentation is performed either at the mid point of the phone or at the point of least change.

These units cannot be recorded in isolation, so the problem with any unit has always been how to excise it from running speech. Coarticulation effects blur the boundaries between units, and this is the reason for choosing diphones or half-syllables. With these units the captured acoustic signal becomes, in effect, a model of the boundary between units. There are some serious problems with this approach.

That speech is made up of concatenated sounds–the main theory is classical phonetics–is an idea deriving from theories which are based on how human beings *perceive* speech, not necessarily how they actually produce a speech waveform. It is clear that individual speech sounds have some kind of psychological reality; that is, they have a valid representation in the minds of speakers and listeners. But it is not clear that such isolated units percolate all the way down through the speech production process, and in our view insufficient empirical evidence has so far been produced to support this model.

But even if the 'segment + coarticulation' model is appropriate, contextual effects from the juxtaposition of the proposed segments can extend to at least four or possibly five

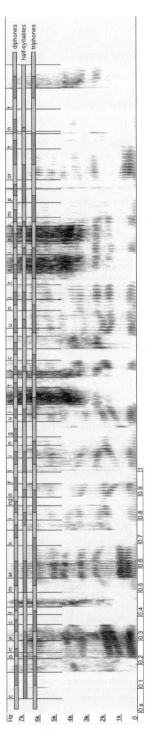

Figure 2.3 Spectrogram showing marking for diphones, half-syllables and triphones on the utterance *Correct marking of linguistic units is important*. Ordinary orthography is used for labelling units to avoid theoretical issues concerning levels of abstraction: the emphasis is on the positioning of the various boundaries. Note that the aspiration phases of syllable-initial voiceless plosives are treated as distinct elements.

segments along the line (in both directions: forwards and backwards). Producing an inventory of all possible such combinations would be very difficult if not impossible.

Even in those systems whose database consists of diphones, triphones or syllables, it is still necessary to mark phone boundaries within those segments designed to take in these boundaries. The reason for this is that subsequent processing may need to modify, for example, the duration of individual segments dependent on the final rhythm–as with the slowing-down effect at the end of sentences or some phrases. An example of this occurs later when we discuss how intonation is rendered in SPRUCE (Chapter 32).

The precise positioning of the boundary is a problem since the boundary notion of classical phonetics derives from a more abstract representation of speech than the physical models of concatenative synthesis. Put another way, the idea of segments and segment boundaries does not stem from inspection of the soundwave or the articulation which created that soundwave; it stems from earlier more subjective ideas based on perception. There is a clear mismatch here because the driving theory of concatenative synthesis is based on the separate segment construct and this position forces us to view the soundwave *as though it were* a blended sequence of separate phones. Furthermore, the theory which deals with how the soundwave is processed in running speech also incorporates this idea. Take the example just alluded to: a speaker slows down the rate of delivery toward the end of a sentence. This translates into strategies for prolonging the duration of individual segments or segments within syllables. Different phones, it is said, behave in different ways: a 10% increase in the rate of delivery of the entire utterance does not translate into a 10% decrease in the duration of each phone within the utterance. Phone-based theories find it hard to state clearly the circumstances under which different phones behave differently.

When dealing with the theory we see that the perception of a string of sound segments does not presuppose a string of sound segments in the actual acoustic waveform. The interim explanation that the apparently continuous waveform is just these segments blended together or overlapping is untenable in the light of evidence suggesting that there are measurable effects of a segment within four or five segments either way. We believe the basis of these ideas merits a serious rethink. There is very little evidence to suggest that the continuous soundwave derives from an underlying set of isolable segments. But there *is* a considerable body of evidence to suggest that isolable segments have psychological reality. That this should translate directly into a physical reality is as absurd as saying that a rainbow is made up of physically distinct, but overlapping, colours–the notion of distinctness of colours in the rainbow is a cognitive illusion, just as the notion that there are distinct sounds in speech is a cognitive illusion. There are no contradictions here–there is no reason to believe that there must be a direct correlation between cognitive entities and physical entities, and certainly no reason to say that a particular type of cognitive representation *must* translate into a corresponding type in the physical world.

Nevertheless, since perhaps all synthesis systems still rest on this theory, we have to make the best of things in terms of understanding how to build and label a concatenative synthesis database–based on any of the currently popular units.

Representation of Speech in the Database

In concatenative synthesis, the database usually consists of stored waveforms. This is true whether we are dealing with small units such as diphones, or large continuous databases in a unit selection system. Waveform storage guarantees spectral integrity of the signal; and, provided the digitisation was done at reasonable sample rates to a reasonable bit depth, it optimises those elements of perceived naturalness which come from spectral integrity. But apart from this, the waveform could well be stored in a form other than its raw state. Leaving aside various techniques for compressing the raw waveform, we refer here to parametric representations of the type found in what has become known as formant synthesis.

The distinction between formant synthesis and concatenative synthesis mistakenly rests on the type of representation. But the distinguishing property of these two approaches in text-to-speech systems is really whether the representation is of an abstract target for individual segments (formant synthesis) or whether it is of a physical signal (concatenative synthesis). In formant synthesis the representation in the database has to be parametric–or at least is advisedly parametric, since the difficulties of manipulating an actual waveform spectrally are for the moment insurmountable, but in concatenative synthesis the representation could be the raw waveform or it could be parametric. There are two distinct advantages to a parametric representation (both exploited in formant synthesis):

- The data is considerably compressed compared with a raw waveform.
- Manipulation of the signal both spectrally and temporally is easy and quick, since there is usually a separation between the source component and resonance features.

Figure 2.4 represents waveform and spectrogram displays of the utterance *How are you?* In (a) the representation is of a raw signal, and in (b) of the same signal represented parametrically using the eleven parameters of the original Holmes formant synthesiser. The results do sound different, but the signal derived from the parametric representation is not particularly less natural sounding than the raw waveform. The question synthesiser designers would have to ask is whether the gains of compressed representation and signal manipulability outweigh the loss of naturalness.

When it comes to synthesising expressive content in speech there is no known way of manipulating the acoustic signal to include the required subtleties. It has been suggested that one way round this would be to include multiple databases each reflecting a different expressive content. Thus there would be an impatient database, an authoritative database etc. As an interim measure such an approach may have its uses, if only to underline the enormity of the problem of collected databases for expressive synthetic speech. The fact of the matter is that expression in speech is continuously varying and the shades of possible expression and emotion still defy description and classification, let alone are able to be treated in a coherent model (Scherer 1996; Johnstone and Scherer 2000; Ekman 1999; Plutchik 1994). The use of a more manipulable representation does at least enable us to include what little knowledge we do have at the moment of how expression comes across in speech, and at best enables us to manipulate the parameters of expression (the prosodics of speech) in a way that enables us to experiment with the subtleties of expressive content. There is no doubt that improving our approach here is both necessary and urgent if we are to move toward genuinely natural-sounding synthesis.

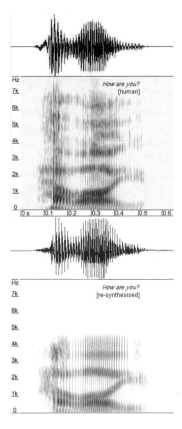

Figure 2.4   Spectrograms of the utterance *How are you?* The top trace was produced by a male
speaker, whereas the bottom trace is a re-synthesised version of the same utterance based
on a parametric analysis of the speaker's raw signal. The analysis was performed fully
automatically and with no retouching, using Holmes' own (unpublished) analyser, and re-
synthesised using his low-level synthesiser. Note that the bandwidth of the re-synthesised utterance
has been held here at 5kHz, which explains the amplitude disparity for fricative consonants
between the two waveforms. The file size of the original mono .wav recording, sampled at 16kHz
with 16-bit amplitude depth, was 14KB, whereas the file of parameter values on which the
re-synthesised version was based had an uncompressed size of just 1.7KB, with all values fully
editable. The file of parameter values had no effect on the bandwidth of the final synthetic
waveform: this was determined by the characteristics of the synthesiser.

Unit Selection Systems: the Data-Driven Approach

Although all practical approaches benefit from the prior existence of a coherent underlying
theory, the unit selection method is not a *theory-driven* approach, but a *data-driven*
approach–there will be experiments to discover what works, rather than develop a theory
which might *explain* what works. They are moving away from previous speech synthesis
techniques toward the kind of pragmatic engineering approach which has enjoyed relative
success over the past couple of decades for automatic speech recognition. There is a
difference, however, between a theory which explains what works and a theory which lends

credibility to the engineering approach–the theory may know nothing of synthesis, but only something about human speech: in this case it would have no chance of explaining what works. There are human speaking strategies which work (by definition) and which are potentially, if not actually, well characterised in the theory of human speech production; there is no reason why such a theory should not be used in underpinning synthesis strategies. Where *direct* usage is not possible it may still help. So for example, acoustic synthesis cannot benefit directly from a model of speech motor control, but as HLSyn (Stevens 2002) has demonstrated, there can be an *indirect* benefit, in this case a formal way of relating perhaps otherwise disparate acoustic effects. HLSyn does not scientifically *explain* what works in the acoustic signal–in fact, in this sense it is rather counter-theoretical–but it does enable a better coherence between acoustic strategies in a low-level parametric system.

Unit Joining

No matter how carefully the cut points between diphones etc. are chosen, there are bound to be amplitude and spectral discontinuities introduced. The designers of the database are careful to cut at points of minimal change, but a careful look at spectrograms shows, for example, that the changes which may be occurring within a phone for one formant are not timed similarly for another, as seen in Figure 2.5. Where there is considerably coarticulatory distortion of target values for the parameters, there is also likely to be severe conjoining errors. Figure 2.6 illustrates the point.

Amplitude smoothing between concatenated segments is usually introduced to try to avoid audible mismatches, though there are a few formal research attempts to determine the thresholds for noticeable mismatches. The situation is almost certain to vary between parameters. The process of smoothing is called *equalisation*; it can be applied beyond amplitude, but in practice this is the only smoothing done. Ideally there would be spectral smoothing also, but with concatenated waveform systems this is very difficult or even impossible. If the units are parametrically represented then smoothing can be systematically applied to each parameter. This might be a tremendous bonus for the parametric approach, but its advantages would have to be offset against a drop in naturalness.

The perceptually disturbing effects of amplitude mismatch between segments varies depending on the type of segment. We have found, for example, that amplitude mismatches in the SPRUCE system, where the minimal units are syllabic in size, are less intrusive than those between phones or diphones. Boundary normalising of amplitude between syllables turns out to be relatively easy employing a simple linear smoothing process related to, say, the initial and final 25% of the duration of the entire syllable. There are two possible approaches with syllables when it comes to boundary smoothing:

- Within the database, or at the moment of extraction from the database, the syllable endpoints can be amplitude-normalised to some standard level, irrespective of the final context. Thus all syllables begin and end with the same amplitude for the same class of phonetic segment.
- The extracted segment is matched with the segments to which it is conjoined. Thus syllables will potentially begin and end with different amplitudes each time they are used.

Strictly, the second amplitude adjustment process should yield more accurate results, though more computation will be involved. Perhaps a combination of the two approaches will yield

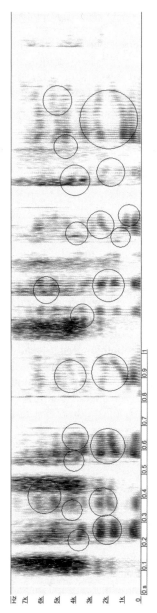

Figure 2.5 Spectrogram of the phrase *Steady state points are difficult to find*, illustrating the difficulty of locating points for cutting diphones in running speech. The circles highlight areas of difficulty, particularly where coarticulation effects are not symmetrical across different formants, or where because of continuous formant 'movement' there is no steady-state point at all. Sometimes formant targets for individual sounds are missed altogether because of the coarticulatory pressures of segments on either side of a particular sound.

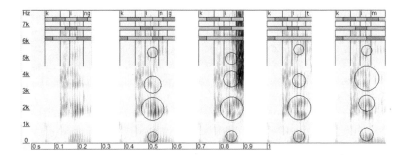

Figure 2.6 Spectrograms showing conjoining errors (circled) occurring when the [ki] excised from the word king–the first spectrogram–is joined to other segments, respectively: [n g], [s], [t] and [m].

the best results. In the first approach the normalised levels differentiate between classes of continuant segment:

- vowels
- liquids and semi-vowels with periodic source
- fricatives and affricates with periodic source
- fricatives with aperiodic source.

The reason for recognising these separate type is the differing intrinsic amplitude associated with different source and filter types. The rules for determining the standard level or the rules for concatenating syllables in the second approach begin by assuming that the segment type at the end of the first syllable is the same as the segment type at the start of the second syllable. Rules refining the procedure when these segments are of different types ensure better smoothing.

Cost Evaluation in Unit Selection Systems

Unit selection systems introduce two kinds of cost notions: target cost and join cost.

- *Target cost* is how close a database unit is to the required unit from the utterance plan.
- *Join cost* is how well two units to be concatenated actually joint together.

The object of evaluating the cost function is to minimise both these costs. In practice this might mean optimising rather than minimising because there is likely to be no perfect solution to either cost.

Prosody and Concatenative Systems

One of the main problems with concatenation systems is the way excised elements preserve their original prosodic concomitants–fundamental frequency, timing, amplitude–or rather the continuously changing values of these attributes within the excised segments. It is almost guaranteed that target utterances to be produced by the synthesiser will need these attributes to be specified differently from their original values. In general, concatenation systems are able to use quite successfully the segmental properties of the original waveform, but much less successfully its prosodic properties. Indeed the prosodic properties will almost certainly need modifying.

The segmental properties, however, may sometimes–though much less frequently–need modifying too, and often this will be as a result of prosodic modifications. An example of this will occur when timing is modified beyond a certain threshold in such a way that the predicted boundary functions–coarticulation, for example–do not behave in the expected way. This occurs in human speech production during slow or fast speech, when coarticulatory processes are either minimised or maximised. If the speech synthesis database is based on an average rate of utterance delivery, these departures will not be represented and the resultant speech will sound unnatural. Modifying the waveform with respect to these changes in the effect of coarticulation involves altering things like formant bending–practically impossible with the raw waveform, but more easily accomplished if the waveform has been parameterised to enable manipulation of individual parameters such as formant frequency. The point being made here is that there is very often an interaction because of the demands of a *changing* prosody–the normal situation–and the spectral content fitted within the prosody.

Prosody Implementation in Unit Concatenation Systems

Prudon *et al.* (2002) have a fairly novel and interesting approach to prosody generation in a small-scale unit concatenation synthesis system based on diphones. In addition to setting up the diphone inventory they also record a database or corpus from which they derive symbolic representations of its prosody and 'phonemic information'. These abstract contours are later used as the basis for adding prosody to novel utterances composed of diphone strings.

Their database consists of 2 hours of speech, annotated using

- normal orthography
- phonological/phonetic transcription
- intonation (called 'pitch') contours
- phoneme/phone information.

Despite a few terminological problems (undifferentiated use of phonological and phonetic transcription, intonation curves referred to as 'pitch contours', and ambiguous phoneme *vs* phone information), it is clear that their initial approach is the markup of the database with a multi-tier symbolic representation. They then transfer the symbolic markup of intonation (which exists independently of the segmental markup) to the MBROLA (Dutoit *et al.* 1996) system, resulting in the fitting of diphone sequences from novel utterances to these existing intonation contours.

Prudon *et al.*'s report of the research stops just short of explicitly explaining whether they think they are fitting prosody to segmental representations, or segmental representations to prosody, but we can imagine here the beginnings of a system which might eventually regard prosody as dominant. The group report that results are not up to the best rule-based systems for deriving prosody, but offset this against ease of incorporating different speakers' prosody and the removal of the need for syntactic processing in text to speech applications of the system.

2.1.4 Hybrid System Approaches to Speech Synthesis

Hertz (2002) recognises that formant-based low-level synthesis and the concatenated wave-form approach need not be mutually exclusive, and proposes a hybrid approach. Her idea is to produce stretches of waveform using a set of formant-based rules and concatenate these with natural speech waveform stretches taken from a pre-recorded database of natural speech. The claim is that a hybrid system of this kind produces more natural-sounding speech than can be expected from an entirely rule-based system. The fact that some segments can be perfectly well generated 'on the fly' suggests that the concatenated waveform part of the system can rely on a smaller than usual database. Such as system may well prove useful in highlighting the strengths and weaknesses of the two approaches by providing direct comparison on a common platform–something that is rare when systems are being compared.

3

Text-To-Speech

3.1 Methods

For the greatest generality–and therefore maximum usefulness–a text-to-speech system should be able to read any text out loud. In practice there are limitations, and although there are many systems available which aspire to this goal they are often less successful than systems which confine themselves to limited domain or niche applications. The reason for this is that there are aspects of text-to-speech which are as yet only incompletely modelled, and systems which need only to access a minimum of these aspects tend to sound better. Speech synthesis, as opposed to the direct replay of recordings or reordered parts of recordings, is called for in applications where creating novel utterances is important.

The distinction we make between recordings and synthesis is perhaps a little narrow because there is an argument for encoding recordings using some of the techniques available in speech synthesis. The argument hinges on storage and access, and is worth developing, because techniques like resynthesised speech or copy synthesis might actually prove to be useful research tools as well as enabling storage of pre-recorded speech in compressed form. A well designed parametric analyser, complemented by a suitable synthesiser, enables very useful compression. Figure 3.1 shows two spectrograms: (a) is derived from a raw waveform of the sentence *Resynthesised speech can be used as a compression tool*, and (b) is the waveform of the same recording after analysis-by-synthesis. The analysis used the eleven parameters of the Holmes system (1964) with the waveform file occupying more bytes than the parameterised file.

3.2 The Syntactic Parse

In most text-to-speech systems a syntactic parse forms the basis of the prosodic component. This is because it is felt by most researchers that prosody revolves around stretches of an utterance which are delimited by certain syntactic features. There are as many different approaches as there are text-to-speech systems, but notable among automatic systems for analysing large databases is the classification and regression tree (CART) approach developed by Hirschberg (1991). Statistical and semi-statistical techniques are often used, including expert systems (Traber 1993), but their results as syntactic parsers are less than ideal because they tend to confine themselves to an analysis of the surface structure of potential utterances.

In any case there remains the practical question as to whether what is achieved by the wide variety of parsers in use is in fact sufficient. But an argument about the sufficiency of

Developments in Speech Synthesis Mark Tatham and Katherine Morton
© 2005 John Wiley & Sons, Ltd.   ISBN: 0-470-85538-X

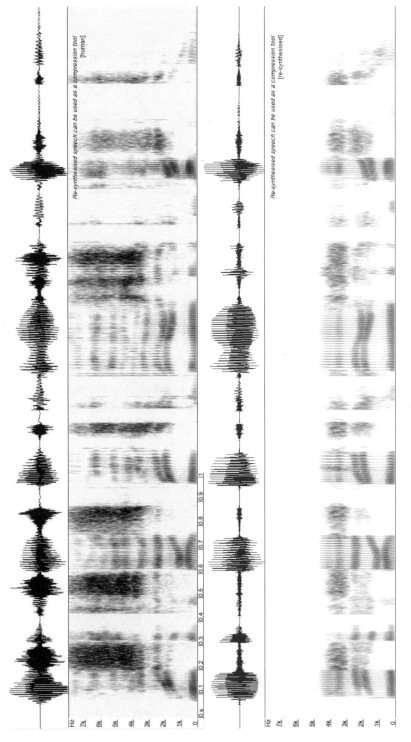

Figure 3.1 Spectrograms of the utterance *Re-synthesised speech can be used as a compression tool*. The top trace is of human speech, whereas the bottom trace is a re-synthesised version of the same signal–see Figure 2.4 for details. The original .wav file occupied 60KB, and the parameter file from which the Holmes synthesiser generated the waveform was just 4.5KB.

this or that parser is irrelevant unless there is good reason for departing from our understanding of what goes on in a human being. It is the usual question: Do we want to try to do our synthesis the way we understand human beings speak, or do we just want to try to produce acceptable synthetic speech without attempting to simulate human processes? It is highly unlikely that statistical or semi-statistical methods are used by human beings except in conjunction with something more along the lines of a linguistics analysis.

But it is perhaps just as unlikely that human beings resort to a full-scale syntactic analysis as is only possible with the full armoury of theoretical linguistics. Here the argument is the same as saying that listeners do not muse on the alternative deep structures of *Flying planes can be dangerous* or similar ambiguous sentences. The analysis might well be done here (if it is done at all) following a probability-based filter which goes for the most probable reading without actually performing all the alternative analyses possible. We do not know what human beings do, but this does not mean that we do not have some idea of what they probably do not do. To reiterate: what human beings probably do *not* do is

- call for a full-scale parse as described by theoretical linguistics
- extract a parse based only on some statistical method other than a simple probability-based filter.

Whatever technique is used in parsing the input text, boundary and other features can be added to the text or its phonological re-representation using rules of prosody. These features form the basis of the utterance's eventual intonation and rhythm properties.

We discuss the assignment of prosody elsewhere (see Part IV Chapter 1 and Part IX Chapter 1), but it is appropriate here to mention *rhythm*. Most researchers in speech synthesis have seen rhythm as a consequence of the timing of the various individual sounds which make up the final output, with little regard to the hierarchically organised relationship between these sounds. It is usually argued that each sound has an intrinsic duration (that is, has a kind of target or abstract duration) which is rendered according to various linear (i.e. non-hierarchically organised) context-sensitive rules to produce a final duration for each of them (Klatt 1979). String these together using a kind of coarticulation algorithm and the final rhythm of the utterance emerges. Rhythm is seen in this model as a cover term for the result of these temporal rendering processes, usually called something like 'duration rules'. Entering into these rules is a consideration not just of the target durations for each segment and some coarticulatory effects consequent on they way they are concatenated, but also the elementary phrase-oriented syntactic structure of the utterance. Hence the general syntactic parse performed prior to assigning prosody assists in the specific assignment of rhythm (*via* the assignment of segment durations) and the specific assignment of intonation and sentence stress (prominence or focus).

4

Different Low-Level Synthesisers: What Can Be Expected?

4.1 The Competing Types

We have mentioned three major categories of low-level synthesiser: articulatory, low bit-rate parametric and concatenated waveform, and have noted some of their differences. But what specifically can be expected of these systems? Table 4.1 shows the main differences when embedded in text-to-speech systems:

- *Articulatory synthesis.* This is potentially good to excellent on all criteria due to the introduction of modelling above the acoustic level. Although the acoustic level is still required, the manipulation of the data is done at the articulatory level which perhaps is more appropriate if human speech production is the basis of the overall model. Articulatory synthesis relies heavily, however, on an understanding of motor control and the dynamics of the vocal tract, and in both these areas there are many gaps in our knowledge–this is the least well modelled tier in the production chain. The preferred basic unit is the gesture, a dynamic representation of a period of articulation; storage might be either in terms of a parametric model of motor control or a parametric representation of the muscular system. One or two researchers (e.g. Browman and Goldstein 1986) use a parametric representation of vocal tract constriction to provide a direct link to the aerodynamics and acoustics of the production system. As with other systems the amount of coarticulatory detail which needs calculating depends on the length of the stored units,

Table 4.1

Synthesis type	Bit rate	Stored units	Database size	Ease of manipulation		Naturalness			
				f0 and time	*Spectrum*	*Segments*		*Prosody*	
						Isolated	*Coarticulated*	*Basic*	*Expressive*
Articulatory	mid	gesture	mid	good	good	good	mid	good	good
Parametric acoustic	low	any	small	good	good	mid	poor	good	good
Concatenated	high	any	large	good	poor	good	good	mid	poor

Developments in Speech Synthesis Mark Tatham and Katherine Morton
© 2005 John Wiley & Sons, Ltd. ISBN: 0-470-85538-X

with the usual inverse correlation between quality and amount of calculation–the best indicator we have, perhaps, that phone + coarticulation is not necessarily the best model.

- *Parametric acoustic synthesis.* What might be excellent synthesis quality is generally marred by poor representation of individual phones and poor rules controlling how they coarticulate. What is gained is the very low bit rate of this type of synthesis, and good rendering of prosody. In theory, parametric acoustic synthesis can be bettered by articulatory synthesis, but this has not happened yet because of the thinness of supporting theory and models of articulation. Parametric acoustic synthesis enables the widest range of manipulation of the acoustic signal which leads straight to excellent prosody and rendering of expression. If expression is sometimes disappointing this is almost always due to inadequacies in the expression model rather than the manipulability of the synthesiser parameters. Hybrid systems, like HLSyn, show some promise, but are really designed to provide an integrating tier over the segment + coarticulation model, rather than bring the force of full articulatory synthesis to bear on the synthesis. The Browman and Goldstein approach is more likely to succeed in the long term, in our view, than hybrid systems, since the articulatory tier in such systems is a no-compromise implementation of a model of motor control and articulation rather than serving simply as an acoustic integrating system.

- *Concatenated waveform synthesis.* We deal with concatenative systems in detail elsewhere, but here it is sufficient perhaps to say that in our view this is an interim technology which is designed for a short-term fix to problems of naturalness experienced by systems which rely on calculated coarticulation and parameter manipulation for good segmental rendering. In principle, since what the listener hears is a recording of human speech, the systems are as good as the quality of the recording; in other words, they should provide an output which is indistinguishable from human speech. Probably one of the main distinguishing features of concatenated waveform synthesis is the fact that the synthesiser is relatively unaware of the characteristics of the data it is processing, whereas to function at all parametric synthesisers must be maximally knowledgeable about their data.

Unfortunately concatenative systems are insufficiently strong in the area of variability–to which human listeners are particularly sensitive. In theory, a stretch of recording can never be correct because it is fixed and invariant. Attempts to manipulate the signal fall short of satisfactory because of problems inherent in current signal processing techniques, like the probable impossibility of satisfactorily manipulating the detail of the signal's spectral content. Unfortunately researchers seem to have underestimated the extent to which the signal does have to be manipulated–even before the problems of prosody arise. The sophisticated use of prosody to capture expressive content is still beyond the reach of any synthesis system because of the impoverished supporting models, but we do know enough to know that it is unlikely that the moment-by-moment manipulation of timing and spectral content will prove adequate. For now, reliance on multiple databases for introducing different styles to the speech output is possibly no more than a crude stop-gap. Having said that, there is a very positive gain to be had when attempts are made to manipulate the waveform. The research is especially revealing of those areas of the signal which need manipulating and which therefore are the carriers of expressive content.

4.2 The Theoretical Limits

In terms of the theoretical limits of the various systems it is fairly clear that articulatory synthesis, provided it covers both the motor control of muscles and vocal tract articulation (i.e. the movement of the articulations as well as the muscles that control them) has the greatest potential. Like all approaches, articulatory synthesis will be as good as the model which underlies it, and potentially this has the best chance at near-perfect synthesis because it takes in the possibility of a complete model of speech production in humans.

Parametric acoustic synthesis will be as good as the acoustic model, and relies heavily on the choice of parameters, from the minimalist approach of Holmes (Holmes *et al.* 1964) to the maximalist approach of Klatt (Klatt and Klatt 1990). However, more parameters does not necessarily mean better results. Parametric synthesis in the form of *re*-synthesis (i.e. not text-to-speech synthesis, but analysis followed by re-synthesis of the extracted parameters) can be very successful indeed from the point of view of naturalness. The fact that resynthesised speech can capture even some of the finer points of expressive content does mean that such systems are adequately manipulable. The problem, however, is that we do not have a good enough model of the acoustics of expression to exploit the possibility fully yet. Because it is parametric this type of synthesis can never be as good as a well-recorded waveform–that would require an infinite number of parameters.

Concatenated waveform synthesis is in theory perfect in the special case that a particular utterance is a complete replay of a complete recording. But this would be self-defeating. Under certain circumstances waveform synthesis sounds very natural indeed, particularly when the units used are long or carefully selected variable units (see unit selection, Chapter 2). The problem arises when the required speech departs more than minimally from the original recording(s). The greater the distance between what is in the database (albeit rearranged) and the required output, the greater the error and the less natural the output sounds. A watershed is reached when changes of prosody or expressive content are needed. Here the system breaks down with respect to naturalness–not, as in the general case, because we do not understand enough about modelling the acoustics of expression, but because the required signal processing is not yet up to the job.

4.3 Upcoming Approaches

We have mentioned several times the hybrid approach of some low-level synthesisers which attempt to manage the acoustics by the introduction of an overseeing tier based on articulator movement. These systems promise limited improvements because they do provide an explanatory integration of some of the detail of the acoustics–obviating the need for its calculation on a hit-and-miss basis.

True articulatory synthesis is a long-term venture because of the poverty of existing models of speech production at the neurophysiological level. As such models become available, the articulatory approach may well come into its own; the addition of layers of computation are of no consequence these days–contrasting sharply with the problems met by pioneers in the area (Werner and Haggard 1969; Tatham 1970a). Improved and properly hierarchical models of the acoustics of speech production will probably reduce any role of an overarching articulatory manager, leaving the way open for true articulatory synthesis.

The future of concatenated waveform synthesis lies with multiple databases to cover a range of different voices and a range of styles or expressive content. Clearly, though, there is a very practical limit to storing and accessing multiple databases, to say nothing of the difficulties of recording large and multiple databases. One possibility for multiple databases is the straightforward transformation of one database into another (say, changing a woman's voice into a man's); if successful this would mean that the preparation of these would not be such a daunting prospect. Although there have been one or two very limited demonstrations of this, there is in fact a logical error here–if you can transform an entire database in advance, you might as well save storage and do it on demand and 'on the fly'. In our view, concatenated waveform synthesis will need some major as yet undefined breakthrough to go much further than the current state of the art as demonstrated in the laboratory. However, for niche applications the current approach will be pushed a little further with rewarding success.

5

Low-Level Synthesis Potential

5.1 The Input to Low-Level Synthesis

Low-level systems which form part of a text-to-speech process normally receive several different types of input. In our terminology, the composite input data stream constitutes the utterance plan. The utterance plan is the interface representation between high- and low-level synthesis; it includes

- a string of *surface phonological objects* (in phonetics: *extrinsic allophones*) whose derivation stems ultimately from the original orthographic text input re-represented as a string of underlying phonological elements;
- *basic prosodic markers or tags* indicating word and sentence (prominence or focus) stressing, rhythm and intonation;
- (in recent and future systems) *sophisticated prosodic markers* indicating initially general expressive content, and later, subtle dynamic expressive content.

The segmental objects and the various prosodic markers or tags are highly abstract and bear only a nonlinear relationship to the rendered acoustic signal that the low-level synthesiser must develop. In the linguistic and psychological theory that defines these input types, the focus is centred on the *psychology* of speech–how it is planned by the speaker and how it is perceived by the listener. Perhaps unfortunately for the speech technology engineer, speakers' and listeners' representations are symbolic and not *directly* able to be transformed into the acoustic signal (in the case of synthesis) or derived from it (in the case of automatic speech recognition). Even classical phonetics, responsible for developing transcription systems like the International Phonetic Alphabet, are based on a listener's subjective perception of articulations and their accompanying acoustic signals, *not* on their objective characteristics. The biggest single problem that has to be faced here is the psychological reality of the segment which tricks people into believing that the segmental structure of speech which they have in their minds carries over somehow directly into the acoustic signal. It does not.

The task for low-level synthesis in text-to-speech systems is to *render* the abstract plan delivered by high-level synthesis to produce a corresponding correct or appropriate sound output. Because of the inherent nonlinearity of the corresponding system in human beings, the synthesis task is far from straightforward. *Rendering* is a term adapted from computer graphics (Tatham and Morton 2003) where the analogy might be with the process which adds colour, texture and reflections to a simple underlying wire frame representation of the

Developments in Speech Synthesis   Mark Tatham and Katherine Morton
© 2005 John Wiley & Sons, Ltd.   ISBN: 0-470-85538-X

graphic. The wire frame is equivalent to the objects of the utterance plan embedded within an overall structure–the shape of the wire frame and therefore the object it represents (even though it is not yet a *picture* of the object). Colour and texture are equivalent to markers associated with the wire frame representation: *When you render this frame make the derived object red and smooth.* Reflections are more dynamic and are to do with the *environment* in which the object is rendered: expression is like this in speech. A person is thoughtful, for example, and this becomes the overall expressive environment within which their utterance is rendered. No analogy is perfect; it is a pointer toward understanding. But for us the concept of rendering is far more productive than simple *realisation* or *interpretation*–earlier terms basically defined within a non-dynamic or static view of speech production and perception.

5.2 Text Marking

The marking of the utterance plan would, in an ideal world, take place entirely within the high-level system. But because of the fact that text barely encodes any of the underlying prosodic and expressive features needed for satisfactory low-level rendering, these need to be added as part of high-level synthesis processing.

5.2.1 Unmarked Text

We know what unmarked text is: it is the kind of text in this book. It is text that contains virtually nothing to assist either the human reader or a text-to-speech system in determining the environment within which it is to be rendered as an audio signal. The environment here is the prosodic or pragmatic environment which needs to be supplied after the text has been input. The question is whether we can identify existing textual objects which can be marked (hopefully automatically) to assist the synthesis process. There are a number of questions for high-level synthesis, among them:

1 What is it that has been omitted (what features) that we need to reconstruct?
2 What are suitable objects and categories for representing the omitted features?
3 How do these categories relate together to give the overall environment for the underlying segmental structure which *has* been represented in the text?

5.2.2 Marked Text: the Basics

There are different depths of marking; we could call this the marking *resolution*. An obvious basic marking involves the normal text processing found in text-to-speech systems–implementing things like rewriting non-alpha symbols (like numbers), expanding abbreviations, disambiguating some abbreviations.

We can supplement this basic segmental marking with various levels of expressive marking which help establish the environment for speaking the plan derived from the segmental representation. The usual example here is that of a drama script:

Mary [with feeling]: . . . if it's the last thing I do.

or

```
<expression type="with_feeling"> ... if it's the last thing I do.
</expression>
```

In the drama script the actor is prompted *how* to speak the text to be spoken, the actor's *lines*. Using XML notation (see Parts VI and IX) we might tag the actor's line here with the element `<expression>` to mark the environmental expressive *wrapper* for the utterance, and add the attribute `with_feeling` to show that Mary is to put some feeling into how she speaks the line. Notice that the attribute symbol (`with_feeling`) does not say what feeling actually is–just as the text . . . *if it's the last thing I do* does not define the sound-wave: it too is just a symbolic representation. Both the actual text and the symbolic element and attribute need to be rendered. In acting we might speak of the actor's *interpretation* of what the playwright has written. It is exactly the same with a musical score–an abstract symbolic representation of the acoustic signal–to be interpreted by a performer. Actors and performers will give us different interpretations of the same underlying symbolic representation: no two interpretations (we would call them *instantiations*) are ever the same, even from the same actor or performer.

With both the marked text or the annotated musical score (*con espressione*, for example) the marking is insufficient to give the interpreter other than a hint as to what the author or composer was getting at. The process of rendering is an active one which *brings* something to the process, and the same has to be true of the synthesis process. It follows that our expectation of what is usually called *detail* in the markup should not be set too high. We shall see that detail is probably not the right word, because it implies that acoustic-oriented information is lifted from the markup itself–in fact this is probably not the case. The symbolic abstract markup serves as a *trigger* for or *pointer* to prosodic aspects of the dynamic rendering process; the detail comes from that process, not from the input.

There are two other types of markup that we will mention briefly here.

> When a synthesis system is properly embedded in a dialogue system we are unlikely, certainly in the future, to see text at any stage. The phonological shape of the utterances will be derived directly from some *concept* input. This concept input is likely to be a parametric symbolic representation of the semantics and inherent logic of what is to be spoken. The concept representation will be converted to an underlying phonological representation similar to that derived from text in text-to-speech systems.

The conversion may be by lookup, or some statistically based approach like a Bayesian expert system, a hidden Markov device or an artificial neural network. How the conversion is done is not within the purview of speech synthesis, though what is to be included is important to us. The symbolic representation of concepts will include the features which text encodes (but unnecessarily so here) and those features which text does *not* encode. Thus general prosodic and local expressive markup of the utterance will be derived from the concept input rather than by *intervention* (either automatically or by hand) as has to be the case with text input. There is no reason though to make the markup any different, however it is derived. Indeed, for the sake of compatibility it would be a good idea to standardise prosodic and expressive markup for *any* source.

Sometimes in dialogue systems the output of an automatic speech recognition system needs to be re-synthesised. For example, the system may need to repeat back the words it has just heard from the human side of the exchange. But perhaps more frequently the dialogue system will need to respond to the expressive content of what the automatic speech recognition system has heard. What expressive content to choose for the synthetic response is the responsibility of the dialogue system, but how it is to be represented must conform with the requirements of the synthesis system which is to render it. In such systems provision has to be made for compatibility and inter-operability between the output of the automatic speech recognition component and an input to the speech synthesis component *via* the dialogue management control.

Before marking the database waveforms for a unit selection system it is necessary to decide on the units to be used for the markup and at what level of abstraction they fall. The marking is to be done using symbols which are abstractions, and these are to be placed on a waveform representation which is clearly not an abstraction in the same sense.

5.2.3 Waveforms and Segment Boundaries

Waveforms *per se* are not the actual signal, but are a representation of the signal which depends heavily on any signal processing which has been used to arrive at what is on the computer screen. There is a sense then in which the display of a waveform is a low-level abstraction, and this level of abstraction is very different from the level occupied by a *symbolic* representation. But there are also different levels of symbolic representation. However, whenever we have a representation of an object it is clear that we do not have the object itself–that is, waveforms through to the deepest of representations are all abstract, differing only in *degree* of abstraction. Linguistics recognises a subset of levels of possible abstractions depending on the functional properties of these levels within the linguistic structure. We are, of course, speaking primarily of the phonological and phonetic components of the overall linguistic structure.

- *Classical phonetics* refers to objects such as phonemes or allophones and the transcription symbols including diacritics which are used to represent them. Phonemes are at the highest level of abstraction in classical phonetics, and allophones (usually differentiated with diacritics placed on some basic symbol) are at the lowest level. Thus /t/ is a phoneme label on a set of lower level abstractions–allophones such as dental, alveolar, aspirated, unaspirated, unreleased, voiced, affricated, glottalised etc.–the set of [t] allophones (Jones 1950; Cruttenden 2001). The analysis of sounds or their articulation in classical phonetics gives us a flat representation rather than a deep hierarchical one, with the difference in abstraction resting on whether the symbol represents a sound or how that sound is articulated (an allophone) or whether the symbol represents a *class* of sounds (a phoneme). It would be wrong to view the classical phonetics structure as entirely non-hierarchical, though, since it is easy to show that allophonic units could derive neatly by rule from phoneme units; this is only a small step from the object-plus-class-label structure which was the original intention.

- In classical *transformational phonology* (Chomsky and Halle 1968) there are abstract objects at the deepest level of abstraction–underling segments–and at the shallowest level–derived segments. Underlying objects are mostly classificatory and related to the job of differentiating morphemes in the language, whereas derived objects embody (in this early theory) most of the cognitively processed information necessary for a relatively automatic phonetics to proceed through the physical rendering process.

- Many *modern phoneticians* would call objects used in the phonological surface representation *extrinsic allophones*; that is, variants which have been derived cognitively from underlying objects within the set of phonological processes. For example, an underlying /L/ systematically derives two different extrinsic allophones in English: /l$_j$/ (a palatalised version) and /l$_w$/ (a velarised version). Since phoneticians are more concerned with dynamic derivations in the phonetic component, they tend to think of phonology also in semi-dynamic terms. It must be remembered, though, that classical transformational phonology characterises phonological processes without the means for selecting from them in any dynamic processing. Variants of objects or segments which arise during the more physical phonetic component are called *intrinsic allophones*–typically coarticulatory variants. Although the notion of two different types of allophone (those cognitively derived and those physically derived) is present in early classical phonetics writers like Jones (1950), they did not have the rigorous categorisation that most modern phoneticians make.

- Those modern phoneticians who support the ideas of the *cognitive phonetics* school (Morton 1986; Tatham 1986a, 1990) probe deeper into the source of intrinsic allophones and model their derivation in terms of physical coarticulatory processes which are subject to cognitive intervention (Lindblom 1983; Tatham and Morton 2004). Psychologists are used to the notion of cognitive intervention in biological processes (Chapter 23), including motor control, but the idea still has some way to go before it is universally accepted in phonetics. The alternative model introduces processes which cognitive phoneticians see as phonetic at a more abstract level in the phonology/phonetics hierarchy, claiming them to be phonological rather than cognitive phonetic processes.

- Other phoneticians, those who espouse the *action theory* approach (Fowler 1980), move some low-level phonological processes into the phonetic component, assigning a much greater role to phonetic processes when it comes to dynamic processing–that is, processes involving time and timing. One or two derivatives of action theory–including Browman and Goldstein's *articulatory phonology* (Browman and Goldstein 1986), attempt to explicitly combine phonology and phonetics back into the integrated relationship that they held in classical phonetics. Cognitive phonetics and action theory are complementary.

5.2.4 Marking Boundaries on Waveforms: the Alignment Problem

The most significant problem met while annotating waveforms is how to find boundaries. Since the symbolic markup used implies segments and boundaries between them, it is natural to attempt to find stretches of speech waveform which can be delimited to match the segment/boundary concept. Every researcher in phonetics knows, though, that the task is an impossible one–boundaries are very hard to find because the concept of boundaries within a waveform which can match up with our symbols is false. The symbolic representations we have are usually elaborations of transcription systems like the International Phonetic Alphabet, and the idea of transcription is based on the theoretical position that speech consists of conjoined segments. The fact that boundaries between these segments are not obvious was explained by the introduction of *coarticulation theory* (MacNeilage and De Clerk 1969: Öhman 1966; Hardcastle and Hewlett 1999) which models the joins between segments in terms of the inertial properties of the articulators (mechanical inertia) or the corresponding properties of the system's aerodynamics (aerodynamic inertia). The model is summarised as:

segment_string . boundary_constraints ⟶ *continuous_waveform*

Implicit in coarticulation theory is the idea that somehow there are ideal segments (often called *targets*) which get contextually modified when abutted with other segments in a linear string. The degree of deformation at the boundaries is a function of articulator targets for abutted segments and the rate of delivery of the utterance. The context of constraints responsible for deformation is linear, though researchers report a context sensitivity well beyond the immediately adjacent segments.

Transcribing the soundwave is an exercise in symbolic labelling requiring the transcriber to identify the idealised underlying segments and write down a symbol which can be annotated to show different degrees of contextual effect. There is little problem with doing this; the theory is explicit enough and its applications useful enough to make the exercise worthwhile. However, difficulties arise if the transcription task involves temporal alignment–that is, involves marking where the segment begins and ends. The reason is simple: the beginning and ending are in this case abstract concepts which cannot *by definition* be marked on a physical waveform.

In marking up a database for future excision of units in a concatenated waveform system, it is necessary to provide a symbolic representation for a stretch of waveform and also to mark the excision points–that is, identify where what is being represented begins and ends. Delimiting segments is driven by practicalities, of course, rather than theoretical motivation. Various researchers have published explicit criteria for marking points in the waveform and calling them boundaries between segments. For example, Hertz (2002) identifies some particularly difficult places (among others) where rule-of-thumb rather than theoretically motivated decisions have to be made:

1 Voiced transitions between obstruents and sonorants were considered part of the sonorants.
2 Aspirated transitions out of stops were considered part of the stops.
3 Transitions between vowels were divided evenly between the two adjacent vowels.

In the SPRUCE system (Tatham *et al.* 2000) we would make decisions 2 and 3 along slightly different lines:

1 Voiced transitions between obstruents and sonorants or vowels are part of the *second* segment.

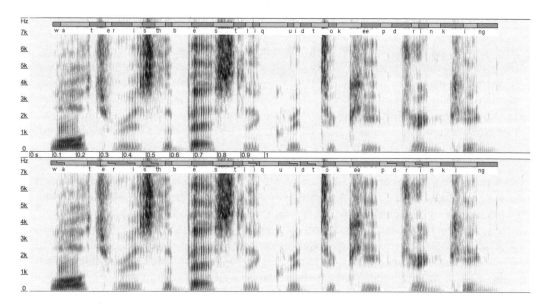

Figure 5.1 Spectrograms showing segmentation of the signal *Water is the best liquid to keep drinking*. The top spectrogram shows segmentation following Hertz's procedures, and the lower one following procedures developed for the SPRUCE project. The main feature to notice is the way in which segments overlap in the lower graph–made possible because of the hierarchical basis of the segmentation algorithm–accounting for apparent telescoping of segments augmenting their coarticulatory behaviour. Some segment boundaries are different between the two approaches: see, in particular, the way in which Hertz assigns the period of aspiration following the release of a voiceless plosive to the plosive itself, whereas SPRUCE assigns it to the following vowel–making the excision cuts occur in different places.

2 Aspirated transitions out of stops are considered part of the following vowel or sonorant–that is, part of the *second* segment.
3 Transitions between vowels are divided evenly between the two adjacent vowels, and then the boundaries for each are found by counting two periodic cycles backward or forward into the adjacent vowel. This is an important point: *telescoped* areas are restored.
4 Transitions between vowels and sonorants are handled using the criteria of 3.

Figure 5.1 shows waveform cutting according to the two different sets of principles.

Most researchers will have their own supporting arguments for adopting this or that procedure. Some will be based on different theoretical approaches, as with 2 above–for Hertz, aspiration is a phase of the stop, but for SPRUCE aspiration is a phase of the following segment exhibiting vocal cord vibration failure. Others will be based on perceptual criteria to be formal, or simply 'what sounds best in the majority of cases'. In 4, for example, we felt that the overall duration of two adjacent vowels is less than the sum of the two vowels in isolation or in different contexts. It therefore becomes necessary to extend the duration of each vowel once it is excised, though they might well be reduced in duration again when re-conjoined in some other context.

5.2.5 Labelling the Database: Segments

The point of this brief summary of different approaches to modelling in phonology and phonetics is to illustrate that there really *is* a problem when it comes to labelling the database waveform in unit selection concatenated waveform synthesis. It may not be necessary actually to name the linguistic theory being adopted; but what is certain is that all of them are very clear on the need to maintain strict separation between different levels of abstraction when it comes to symbolic representation. This translates into a careful avoidance of mixing levels. So, for labelling the waveform what level should be chosen?

- A deep phonological (phonemic) representation where allophonic differences, although systematic, are not included? An example in English would be to label all l-sounds with the same symbol, not differentiating between the phonologically derived 'clear' or 'dark' variants which must be used systematically in certain contexts.
- A surface phonological (extrinsic allophonic) representation where cognitively derived variants such as the two different types of l-sound are noted? This representation is consistent with the definition of the utterance plan to which we keep referring.
- A phonetic representation (intrinsic allophonic) where coarticulatory variants which are not wholly derived cognitively but which arise because of physical constraints in the phonetic rendering process are noted? An example in English would be use of a dental t-sound in a word like *eighth* as opposed to the more frequent alveolar t-sound to be found in words like *top* and *let*.

This list is not exhaustive: there are other identifiable levels which could be used as the basis of the representation. The extreme representation–one which represents all possible acoustic nuances derived no matter where in the hierarchy of processes–cannot be symbolic at all: it *is* the waveform. The choice lies in how we answer the question: *What level of detail is necessary to ensure perceived naturalness?* If it turns out that listeners cannot perceive certain types of detail, then perhaps these need not be represented or play a role in the selection process. Putting this consideration round the other way, the other relevant question is whether there is in existence a recognised symbolic representation which can handle all the necessary detail, or, much more importantly, *whether the required amount of detail can be represented symbolically at all*. This is not so fanciful as might be imagined–it is a real problem in the markup of databases.

We discuss labelling of the database elsewhere (see Part V Chapter 4), but in keeping levels distinct it is important to remember that objects such as phonemes cannot have physical attributes like duration. What they can have is length–an abstract version of duration. But the length is significant in phonemes only if it is contrastive–which it is not, for example, in a language like English, though many languages do have contrastive phonemic length. Extrinsic allophones can have non-contrastive but nevertheless systematic abstract length, along with other non-contrastive attributes acquired during phonological processing. Thus phonologists have vowels occurring as a lengthened extrinsic allophone if they are the nucleus of a syllable whose coda includes a voiced consonant. Physical duration is important in the rendering of prosodic elements like rhythm; but if we are to be consistent we must avoid assigning duration to symbols at the underlying levels.

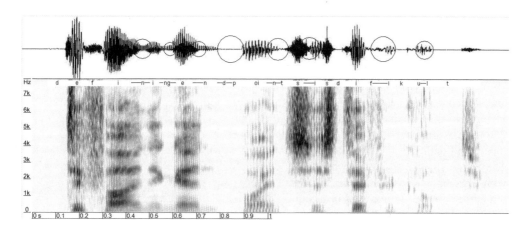

Figure 5.2 Waveform and spectrogram of the sentence *Defining endpoints is difficult*. Circles on the waveform trace show areas of particular difficulty, also indicated in the symbolic labelling by horizontal lines. Note the difficulties surrounding nasal consonants, and abutted plosives where the first of the two plosives is rarely released, but the apparent overlong stop phase is in fact less in duration than the sum of the two target stop phases spoken in different contexts. It is in this type of context that the hierarchical approach allowing for overlap of segments comes into its own (see Figure 5.1).

5.2.6 Labelling the Database: Endpointing and Alignment

This problem of wanting to include in the markup *some* physical properties of the signal, whilst associating them on the one hand with the waveform representation and on the other with a symbolic marker attached to the waveform, presents a very real theoretical dilemma which may have practical consequences. The concept of duration implies start and stop endpoints of a physical object; but the corresponding symbolic label cannot have the same kind of start and stop endpoints.

Consider Figure 5.2, a waveform representation of the utterance *Defining endpoints is difficult*. Any phonetician or synthesis researcher who has had the task of labelling such a waveform recognises how difficult it is to make decisions about where to mark some of the boundaries, while others seem very easy to mark. Thus, where does the *n* in *defining* begin and end? Where is the boundary between the *d* and *p* in *endpoints*? and so on. The point is almost a philosophical one: *n, d* and *p* are not objects which can have endpoints, they are *symbols*. And even if we confine ourselves to the physical level and forget symbols for the moment, since coarticulatory effects (in the phone + coarticulation model of speech) can be found as much as four phones distant from a given phone it just does not make sense to mark boundaries.

But any researcher in synthesis will say that we have no choice: we *must* mark boundaries. And within the theory that most synthesis is conceived under that is true, because the theory tells us that speech consists of objects which do have boundaries. So the marking of boundaries and the measuring of durations all form part of the contemporary approach to synthesis. But this is an artefact of the underlying theory. Change the approach to how speech is structured and these ideas change or even vanish.

We do not wish to be impractical, and certainly would not wish to abandon contemporary theory abruptly, but note that some researchers–the articulatory phonologists–have taken a different approach and split the signal into horizontal time-varying parameters rather than concatenated vertically delimited segments. An early phonological/phonetic theory (Firth 1948) called *prosodic analysis* also used this approach, and our model proposed in Parts IX and X incorporates many of these ideas. An interim formulation, based on an interpretation of Firth's original concept, has also been tried out in a limited way (Tatham and Morton 2003). In essence the idea here is to shift the theoretical focus away from the segment, and thus avoid defining it in the traditional way, and move toward a prosody-centred model. The approach, like that of the articulatory phonologists who were trying to do the same with segments or gestures, uses the continuousness and overarching properties of prosodic rendering to wrap a segmental or gestural *flow* which belies hard boundaries.

Part II

A New Direction for Speech Synthesis

6

A View of Naturalness

6.1 The Naturalness Concept

We are moving toward a generation of synthesisers which changes direction away from the basic reading machine requirement of earlier systems. The change is toward placing more emphasis on natural speech production, moving away from the artificiality (even for a human being) of reading text out loud. This will inevitably mean that the current synthesis model will be less appropriate and tend to become inadequate. The focus is already moving toward the quality of the synthetic voice rather than concentrating on the reading task *per se*. As such, speech synthesis needs to model human speech production, but there are two different, but complimentary, goals to choose from. Is speech to be modelled as

- self-centred behaviour directed toward the goal of producing an acoustic signal which matches up with underlying requirements? or
- behaviour focussed on the goal of causing appropriate responses in listeners by providing suitable acoustic triggers?

In the first of these, where the goal is just to produce a signal reflecting the speaker's planned utterance, modelling is in terms of speech production itself. But in the second it is clear that the process of speech production must be augmented by considerations of perception–and perhaps the perception of the actual listener on this occasion. We believe that speech production is about producing a signal which not only reflects how speech is produced but also constitutes an appropriate trigger for perception.

If we are right in focussing on what has become known as *production for perception*, then naturalness becomes a variable property of speech. We are more likely to require that our synthetic speech be appropriately natural rather than just natural. This is easy to appreciate if we bring into account not just the listener's perceptual strategies but also the general environment in which the speech is to be perceived. Thus:

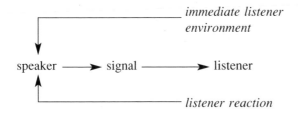

Developments in Speech Synthesis   Mark Tatham and Katherine Morton
© 2005 John Wiley & Sons, Ltd.   ISBN: 0-470-85538-X

In this model the listener's environment and reactions to what he or she perceives are fed back to the speaker who alters the production accordingly, making the signal tailor-made to the circumstances. Theories of production for perception also include a generic perceptual model embedded in the production process to produce speech designed to be perceived where perhaps details of listener and environment are not available. An example of this would be the production of a broadcaster, where neither feedback channel would be available, or a lecturer faced with 20 students. In the second case, though, there would be physical environmental information available to the speaker (e.g. the amount of echo or ambient noise); that is, the environment-to-speaker feedback channel would be available but the listener-to-speaker channel would not.

It is not difficult to see that we may have to alter our ideas about naturalness in speech synthesis. Perhaps this is premature for the moment, but shortly we may have to address not the question *Is this synthetic output natural?* but rather *Is this synthetic output appropriately natural?* We shall be moving inevitably toward an adaptive model for speech synthesis which will be able to simulate the observed adaptation of human speech as environment, listener needs, utterance difficulty and speaker difficulty vary. We have to move away ultimately from the narrow concept of a text-to-speech system–a system that talks out written information–to the wider concept of a device for delivering information without recourse to a written interface. Once we attempt this it becomes untenable to accept constraints which do not allow us to take into account the environmental surroundings available to human speakers.

For the moment most dialogue research does at least take some account of concept-to-speech synthesis, even if *pro tem* an interim text level has to be incorporated. Many researchers have spoken of how the concept includes much more information than text: this kind of observation was made years ago. The reason why synthesis systems have not yet moved their mainstream into this is that there has been little demand; the designers of dialogue systems are not quite there yet in terms of the enabling dialogue technology. Currently computational linguists and others working in the area of dialogue or, for example, translation, consider speech output/input to be the least of their problems. We can, though, start building the kind of 'speech as primary medium' models which will probably dominate in the future. Text-to-speech will continue to have its place because there is a lot of written material out there which needs reading out, but the genuine speech medium will grow up alongside this older and necessarily restricted technology.

6.2 Switchable Databases for Concatenative Synthesis

The concept of switchable databases arises as a result of the need to have different voices and, within one voice, different styles or expressions. There are two main ways of achieving variant speech:

- perform 'on-the-fly' transformations of material from an existing database
- change database on demand.

When human speakers change, everything about their voices changes, and it makes sense, when switching between speakers within a single synthesis system, to call on an alternative database which must by definition have all the inherent properties of the second voice. However, when a single human speaker changes style or comes across with a different expressive content, the change to the voice arises only from the altered style or expression; perhaps almost all of the properties inherent in the speaker's voice remain. Under these conditions it makes less sense to switch databases than to modify the existing speech data to suit the altered circumstances.

Whilst there is little wrong with the logic that prompts two different approaches for different types of voice change, in practice the situation is much more difficult. The difficulty lies in trying to transform an existing style for one voice (the style the database was recorded in) into a different style for the same voice. There are two reasons for this:

- We do not understand the basis of the transformation. Another way of saying this is that our models of expressive content in speech are inadequate. This means we do not have a reliable basis for making alterations to the recorded acoustic.
- Even if we had an adequate model it is questionable whether our current signal processing techniques are up to modifying the recorded waveform. Some parameters can be modified with relative ease and success–fundamental frequency and timing are two obvious parameters. Other features like spectral content are much more difficult and often impossible to modify.

The interim fallback technology to make up for inadequacies in the model of human speech and in current signal processing techniques is to record multiple databases–perhaps of the same material, but certainly by the same speaker under identical recording conditions. The recording environments need to be identical so that elements from one database can be freely mixed with elements from another database. For all but the simplest of material and the smallest number of alternative databases, the recording problem is enormous. Every recording engineer knows that to try to replicate identical conditions on different occasions is virtually impossible. There is the additional problem that human listeners seem to be very sensitive to any lack of coherence in what they hear, and failure of this kind of integrity in the running of a reconstructed signal in concatenated waveform synthesis is one of its major problems as a technique in general, let alone trying to mix material from different databases.

There are therefore several reasons why the multiple database solution is inadequate; but having said that, there are clearly some niche applications relying on very small databases which could benefit from variable style derived in this kind of way. One example might be telephone banking, involving limited functions like reading out balances or completing very simple transactions like transfers between accounts. One can imagine one database which has a matter-of-fact and informative expression (for simply delivering information), another which has a patient expressiveness (for explaining something complex to the listener) or another which is mildly dismissive or scolding (for delivering reprimands to customers who have exceeded their borrowing limits). Without these alternative expressions the system would have to use alternative vocabulary to explain its patience, impatience etc., and the result, whilst adequate in a simple content delivery sense, would fall short of real naturalness.

6.3 Prosodic Modifications

This is perhaps not the place to go into too much detail (but see Tatham and Morton 2004), but the simplistic model which confines prosody to changes of fundamental frequency, amplitude and timing has led to problems with concatenated waveform systems. At an abstract level we speak of intonation, stressing and prominence, and rhythm; but there is no linear correlation between these abstract concepts and the fundamental frequency, amplitude and durational parameters of the acoustic signal. Phonological prosody–the abstract version–is comparatively simple, with well documented relationships between its three prosodic features. But phonetic prosody–the physical version–is extremely complex both in the way its parameters relate to the underlying abstract ones, and in the way they interact at the phonetic level itself.

These facts alone should make the speech synthesis researcher more cautious than has often been the case when they have been misled into thinking that the relationship between phonology and phonetics is straightforward. But it is also the case that varying the obvious physical features of fundamental frequency, amplitude and timing causes other changes, notably spectral ones which we perhaps would not model as prosodic features but which nevertheless contribute to the overall integrity of prosody in the signal. The most obvious example is the way coarticulation alters the moment-by-moment spectral content of the signal. Coarticulation theorists maintain that the phenomenon is time-governed since it derives mainly from the inertial properties of the mechanics and aerodynamics of the vocal tract. So alter the timing (as you would in changing prosody) and you alter coarticulation and therefore spectral content. This is no problem in the human being; but if a synthesis system has a single database with material recorded, therefore, at a single tempo, then there is a need in changing the tempo to change the spectral content also–and this is very difficult with dealing with the raw waveform. Elsewhere (see Part IV) we have suggested that a parametric representation designed to permit this kind of signal manipulation would perhaps be more suitable for storing the database.

Prosody is often taken to be the vehicle for expressive content in human speech. This is a simplistic model, but by and large the features of phonetic prosody are those which change as expression changes. Hence arises the feeling that manipulating expression is just a matter of subtly varying these features. But once again we hit hitherto unforeseen problems when we come to consider what actually goes on the acoustic signal when people speak with different expressive content. Take the very simple example of anger. When a person is angry there are relatively uncontrolled changes to the vocal tract due to increased muscle tension. This results in dimensions to the tract which are different from when the speaker is relaxed, and what follows from this is that there are spectral changes again and, even more importantly, changes to the way in which coarticulation operates as a result of 'stiffness' of the vocal organs. Unfortunately we know almost nothing of the acoustic details of expressive speech and even less of the articulatory details. We can, however, guess enough to know that building expression into the acoustic signal is likely to defeat attempts using current technology to convincingly transform existing databases.

7

Physical Parameters and Abstract Information Channels

7.1 Limitations in the Theory and Scope of Speech Synthesis

One possible model of prosody enumerates the physical parameters available for conveying prosody–these are the parameters of phonetic prosody–and the information channels which cause them to be modified.

The physical properties we are able to change are principally:

- *fundamental frequency*–involves modification in the human being of the rate of vibration of the vocal cords (in synthesis, the repetition rate of the periodic source);
- *segmental timing*–modification of the timing or durational features of individual segments or syllabic groupings which contribute to rhythm;
- *voice source properties*–changes to the spectrum of the voice (e.g. breathy voice, angry voice) or to aspects of its timing (e.g. creaky voice);
- *spectral properties*–modification of spectral tilt (e.g. anger) or spectral composition (e.g. changes in coarticulatory properties of the acoustic stream).

These physical properties are changed by the processes which introduce prosody into the acoustic signal. The information channels which contribute to deciding what to do with the physical vehicles are:

- *segment stream*–the segmental detail of the utterance plan comprising the final encoding resulting from processes performed on the underlying phonological representation (the stream is of extrinsic allophones);
- *prosodic environment*–the basic prosodic wrapper within which the segmental stream is to be rendered;
- *expressive content*–short-term wrapper to determine the required expression of the utterance stream;
- *style environment*–long-term wrapper to indicate the overall style of the utterance.

These are the sources of information which constitutes the material on which the processes of prosodic rendering operate. Bearing in mind that it is usually wise to keep processes and objects transparently separate in any kind of model, let us make some observations about them.

Developments in Speech Synthesis   Mark Tatham and Katherine Morton
© 2005 John Wiley & Sons, Ltd.   ISBN: 0-470-85538-X

7.1.1 Distinguishing Between Physical and Cognitive Processes

There is the obvious need to distinguish carefully between physical and cognitive processes. The reason for this is that they operate in different worlds–physical and abstract respectively –where our premises and expectations differ. One way of avoiding confusion is to use different terms for correlating objects on which they operate. Thus we speak of voice and intonation in the cognitive part of the model, but periodic excitation source and fundamental frequency change in the physical part of the model for correlating concepts. The way in which cognitive and physical objects and processes behave is vastly different despite clear correlating relationships. It is a serious mistake in characterising the correlation to assume one-to-one relationships–this point cannot be overemphasised.

Despite the theoretical dangers we cannot avoid having to say something about how these cognitive and physical processes relate on either a correlating or sometimes causal basis. We clearly have need to characterise exhaustively the acoustic or physical processes in general–not because they have been identified as being used in this or that operation in particular, but so that we can have a complete description should we need at some point to understand detailed properties of the acoustic signal. The comprehensive characterisation of acoustic processes is essential in explaining the properties of the derived acoustic object. But current descriptive techniques are not explicit, and introduce variability arising from inconsistency by the person assigning the label.

We need also to identify and understand cognitive processes since these account for many of our decisions and choices in speech. Both physical and cognitive processes will have to be expressed in terms of some degree of fineness or resolution–processes have degrees of detail just as objects have degrees of detail. Our reasons for settling on particular levels of detail must be explicit, and the results must be compatible between the cognitive and physical domains if correlation is to mean anything. Observing these ground rules enables us to ask questions like these:

- To what extent are the spectral properties of the speech waveform 'used' in speech, and what are the constraints on their usage? The properties of an acoustic signal extend beyond what is actually of interest in speech–but how?
- What about segments etc.? Are our lists exhaustive and appropriate? Note that segments are abstract objects and there is no external reason why correlating physical objects should appear in the signal. Perhaps it would be helpful if they did, but we cannot assume this, whatever we feel to be the case.
- How fine does the degree of categorisation or resolution need to be? . . . and why? Allied to this is the worry that we might miss something if the resolution is not fine enough, but we do not want to be in the position of not seeing the wood for the trees.
- What about the overall constraints of production replication and perceptual discrimination– and are either of these fixed or are they context-dependent? The physical production system has its biological, mechanical, aerodynamic and acoustic constraints, but how do these relate to constraints on perceptual discrimination of the soundwave?

There are many questions of this type which have not been properly addressed in speech theory, or which have been addressed in such a way that misleading answers have been arrived at. This is not an attack on existing theory or methodology; it is more an observation that there must be coherent, explicit and agreed rules for pursuing the science if we are to raise the level of its productivity. In more technical terms, there is a great need to rehearse the metatheory, if only to evaluate the appropriateness of competing models.

7.1.2 Relationship Between Physical and Cognitive Objects

There is the problem of the linearity of the relationship between physical and cognitive objects. We need to ask ourselves whether it is possible to identify unique abstract and physical relationships. For example:

- Is this abstract unit always related to that physical object?
- Are there comparable types of object within the two domains?
- Are there groupings among objects: is this type of physical object always related to this type of abstract object, and in what way?

7.1.3 Implications

The above are very important theoretical questions to be considered within the framework of speech synthesis. Take, for example, the single property of style. There are all sorts of issues here. Are we safe in relating physical phonation styles (breathy voice, creak etc.) with abstract cognitive styles of speaking (broadcasting, lecturing, intimate etc.). The relationship is by no means obvious. Some abstract styles might be associated with rate of delivery, amplitude, voice quality (phonation type), others with spectral content, fundamental frequency contour, and so on. When we get to the point in our synthesis that we need this level of detail in the output, we are some way beyond the considerations normally highlighted in linguistics and phonetics (Black and Taylor 1994), and are into areas of intersection of linguistics between disciplines such as psychology and sociology, to say nothing of biology and neurophysiology. Style, mood, personality etc. all need proper explanatory characterisation if we are to make sense of incorporating them into our synthesis systems. To take just one: how personality is expressed in speech is well beyond current considerations.

7.2 Intonation Contours from the Original Database

There are two different strategies for defining the fundamental frequency contour in speech from a unit selection concatenated waveform system:

1 Derive an abstract phonological intonation contour for the utterance in question. Then

- convert this to a fundamental frequency contour, and
- apply signal processing to the concatenated units pulled from the database to match the computed contour.

–The database size is constrained by how many different units are needed and how many repetitions of each is considered desirable for optimal naturalness of segment rendering. The database is moderate in size. The final contour is idealised (and therefore unnatural) and devoid of variation.

2 Derive an abstract phonological intonation contour for the utterance. Then

- convert this to a fundamental frequency contour tied to the utterance plan (the extrinsic allophonic string to be spoken)
- retrieve units from the database which match the segmental plan and fundamental frequency requirements.

–The database size is constrained by the number of units required and by how many pitch versions of each is considered optimal. The database is huge. The final contour is idealised (and therefore unnatural) and devoid of variation, other than that induced by gaps in the database.

In both cases the utterance plan–the required string of segments–is derived first. The fundamental frequency contour is fitted to the segment string based on how the segmental string has been labelled symbolically for intonation, following a markup system such as ToBI (see Chapter 32).

Klabbers *et al.* (2002) propose an alternative which they feel might improve on both the usual approaches. They propose to mark up the system database in terms of the examples of intonation contours it actually contains. Ideally the markup would be dual (both phonological and phonetic) or hybrid (perhaps what we have called an exemplar instantiation of the abstract phonological markup). But in any case the markup moves the usual phonological abstraction one stage in the direction of an actual speaker–and in this case, importantly, the actual speaker whose voice we shall hear. These curves are excised and new utterances are fitted to them. Their characterisation of intonation is within the rhythmic environment of the utterances found in the database. Our own feeling is that this approach will be shown to be very promising conceptually and from a practical perspective. One or two researchers have found it useful to mark up stressing in the database (e.g. Drullman and Collier 1991; Tatham and Lewis 1992), but few have successfully taken a combined approach to determine the prosodic model.

It is crucial to note which way round the fitting is: Klabbers *et al.* are fitting segmental strings to pre-existing, though idealised, intonational contours. This is halfway toward the proposals we make concerning deriving prosodic contours first and then fitting utterances to them. The Klabbers *et al.* solution is an excellent example of how theory and practicality can converge into a versatile engineering solution to the very difficult problem of adding natural prosody to synthetic speech. Their solution needs further work, and we should like to make the following points:

- The database will have to be large to contain a good set of representative intonation contours, and it will probably be necessary to reduce that set to what we might call distinguishing contours, rather than include a large number of alternative versions of the 'same' contour.
- The database prosody is 'real' and neither contrived nor neutral or idealised. It therefore contains a subset of all possible variations of each contour as it occurs in the real world.
- There is a tradeoff between errors of variation (in this system) and no errors by definition in the two idealised systems above.
- How do errors of variation (picking the wrong one, for example) compare with the unnaturalness of idealised intonation?

- The recorded database prosody correlates with the speaker's expressive content. This may not be appropriate for the utterance to be generated. At best the domain of the database and the domain of the new speech should be the same or very similar.
- They discuss intonation mainly, but there could be difficulties when the other prosodic parameters also change–for example, when the speech slows down. It seems that the phonetic prosodic parameters interact in a continuously variable way, and often interact with segmental makeup such as formant frequency values.

The work Klabbers *et al.* report seems robust enough to be taken further, and this would certainly be a very promising line of enquiry.

7.3 Boundaries in Intonation

Many researchers in the field of intonation begin by dividing text into major word groups occurring within the syntactic sentence domain. In addition it is recognised that there are intonational features which span sentence domains, and this prompts the introduction of the wider paragraph domain. Hence:

```
<text>
        <paragraph>
                <sentence>
                        <major group/*>
                </sentence>
        </paragraph>
</text>
```

Although this XML fragment showing syntax-based groupings includes just one paragraph containing one sentence, there can, of course, be several of either. The asterisk at the end of the major group element signals just such a possibility: 'one or more major groups here'. Punctuation marks may be a rough guide to these groupings, but most researchers who base their intonation model on groups of this kind need a syntactic parse in addition to establish group boundaries. Parses used by different researchers vary considerably in sophistication. For human readers punctuation and text layout seem to be major cues in determining intonation patterning, among other prosodic features such as rhythm. But for the automatic analysis needed in text-to-speech systems, word grouping needs to be as explicit and consistent as possible–hence the need for syntactic parsing.

When considering human reading, Zellner (1994, 1998) draws our attention to findings focussed on the notion of 'equilibrium' as a psycholinguistic principle in determining boundaries. She points out that readers tend to subdivide major word groups into smaller sub-domains–minor word groups–which are fairly well equally distributed with respect to length considered either as number of words or number of syllables. This was a significant finding by Gee and Grosjean (1983). Major groups also behave in a similar way, tending once again toward equilibrium with major boundaries likely to occur at minor boundaries around the middle of long sentences. Zellner observes in her own (1998) data for French that within the sentence domain such a boundary occurs regularly after around 12–15 syllables. Strikingly Zeller suggests 'no association between syntactic factors (for example, the major syntactic break between an NP and a VP) and major pauses of this type–quite the contrary, the syntactic status of preceding and/or following words seems to bear no relation to the presence of an empirically determined group break' (1998). The idea is that as a function of the psychology of language there is some kind of natural or balanced sentence or phrase–one that is in equilibrium–toward which phrases and similar groupings gravitate. Zellner suggests that this human behaviour can be translated into a computational model for arriving at major and minor groupings of words within text.

Although most of Zellner's comments about the intonational structure of speaking and read text refers to data from French, the ideas are in principle extendable to other languages. The main message is that psychological factors beyond plain linguistics kick in under certain circumstances. It may well be that the syntactic structure of the text plays a major role in determining aspects of prosody, but there are other factors too.

There are constraints in the reading out of text which are not inherent in the text itself, but which may themselves become the vehicle for prominence or semantic focus or other phenomcna. This idea is not dissimilar to the important principle in cognitive phonetics, first observed at the segmental level, that 'unwanted' constraints on production can be turned around, elevated in status and used to enhance communication (Morton and Tatham 1980). Zellner hypothesises a basic neutral reading style constrained by these psychological features. The model comprises 'sets of linearly structured minor word groups that are

- regularly interrupted by major word group boundaries
- marked by punctuation, or
- calculated in terms of the "equilibrium" hypothesis'.

There is a clear and noticeable break at the end of the paragraph, which is itself regarded as the basic structural mechanism–the widest domain (in our terminology).

We ourselves avoid the term 'neutral' when it comes to an actually occurring utterance. We argue (see Chapter 11) that synthesising a 'neutral' utterance is not helpful, but that the idea of an abstract neutral utterance which must be rendered within some expressive prosodic context is useful since it forms the basis of modelling prosody separately from the segmental utterance. Zellner's conclusion is important in the development of the theory because it gives us a basis for separating linguistic requirements (part of the utterance plan) from the psychological constraints of the rendering process. This puts psychological and physical constraints on a par–and both are susceptible to cognitive intervention.

8

Variability and System Integrity

8.1 Accent Variation

One source of variation which we may want to include in synthesis systems is dialectal or accent variation. From now on we shall use the term 'accent', because 'dialect' implies variation of syntax and lexis, whereas 'accent' implies only variation in phonology and phonetics. Since in a text-to-speech device the syntax and lexis of the dialect mostly in question are already taken care of, and since the system models only phonology and phonetics, it makes sense to confine a discussion of regional or social variation in pronunciation to a discussion of accent.

There is a claim in the literature that 'post-lexical phonology' often varies with accent. At its most simple this means that once a word is selected the phonological processes it is subjected to can vary. The question to ask is whether lexical selection assumes a generalised phonological representation of the word at the deepest level which is accent-free–the detail of accent variation occurring in phonological processing after the word has been given its underlying representation. If this is the case, then the underlying representation is for the entire language and any accent variations which occur need to be derived from it. The alternative model has a lexical representation which is not the phonological underlying representation and which receives an accent-specific representation as it is selected. So:

- *Accent model 1.* The underlying representation is a highly abstract characterisation of the word and, while being, of course, language-specific is not at the same time accent-specific. Accent variations are derived from this single representation during the course of a general phonology. Rules in the phonology (p-rules) would need to be accent sensitive to arrive at different derivations.

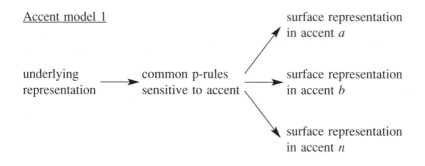

Accent model 1

underlying representation → common p-rules sensitive to accent → surface representation in accent *a* / surface representation in accent *b* / surface representation in accent *n*

Developments in Speech Synthesis   Mark Tatham and Katherine Morton
© 2005 John Wiley & Sons, Ltd.   ISBN: 0-470-85538-X

- *Accent model 2*. There is a lexical representation which identifies the word within the language–call this the *archi-representation*, but also or immediately logically following, provides alternative underlying representations dependent on accent. Each accent would have its own phonology, or there would be a branching phonology so that any one underlying form would follow only a pathway dedicated to its parent accent.

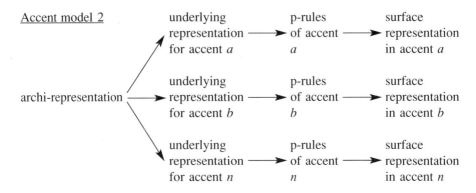

We prefer model 1 if it can adequately account for the observed human behaviour because it captures the most generalisation by having a common set of phonological rules no matter what the accent. Models 1 and 2 would be equivalent *only* if all the phonological rules of model 1 had to be accent-sensitive. Most of them probably do not; so the model immediately shows the commonality between accents. Model 2, on the other hand, not only requires separate sets of rules but requires separate underlying representations.

In our presentation here, both models 1 and 2 ultimately share a common underlying representation–that of the language itself in some very abstract neutral sense. However, most advocates of model 2 do not include the archi-representation. We feel that, without this, the model does not capture the feeling that native speakers have that accents are variants of a single particular language. We do not feel that there is anything distinguishing accents which needs to be captured in separate underlying representations–and if there seems to be, then probably we are dealing with a dialect variation which would involve a separate lexical entry rather than just a variant pronunciation of a single lexical entry. This would be an extreme case, we feel, and every effort should be made in building a model of accent to capture what is common as well as those areas where there are variants. Part of the derivation of three surface accent variants of underlying 't' in model 1 would look like this:

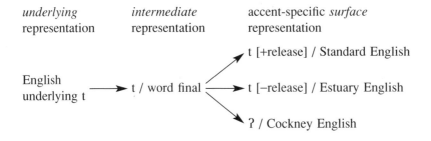

Model 1 enables a compact core characterisation of the language-specific phonological processes without consideration of accent variants. It is also compatible with traditional language- (rather than accent-) oriented phonological models, though it is not the same as these because there may be a need to include accent-specific rules after the usual accent-free surface representation.[1] For this reason we have renamed this the *intermediate representation*, though if we were not interested in accents we would simply stop the derivation process at this point.

We prefer this to the alternative representation under model 2:

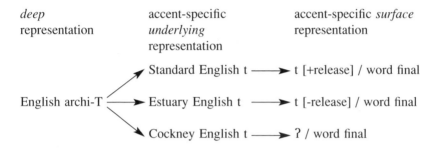

Since in model 2 all phonological rules occur after the accent-specific underlying representation has been derived, it follows that at no point do we have a condensed abstraction of the rules of the language without consideration of accents. In this approach all phonologies characterise specific accents.

This fragment of theory is controversial in linguistics. Traditionally books about applications of theory generally avoid controversy, but we introduce this construct because it is plausible and can help resolve a problem, although this may be a controversial point. But as elsewhere this is because the alternative or earlier theory has perhaps contributed to holding back progress in our field application. The interest in accents in speech synthesis arises because of the emerging need to have devices which can speak in different accents rather than some accent-free neutral speech which is *for that very reason* unnatural in the way it sounds to listeners. Whether a single machine is to speak in different accents or whether it is part of a network of machines performing the same application (for example, regionally based telephone banking), it makes engineering sense to have core code as a universal basis for the application. This enables variant blocks of code (the accent-specific parts of the high-level system) to be compared, and much more importantly be added to as and when necessary without a total rewrite of the system.

For example, current telephone banking generally uses a standard form of the pronunciation of the language. Regional variants can be rolled out without disruption to the existing system with the minimum of effort whether the computer installation is at a single location or geographically distributed. The system would also be fault-tolerant in that any part of the system can take over the function of some other part by falling back on its local accent-free

---

[1] An example of this would occur in some accents of US English where internasal vowels become strongly (as opposed to weakly in the usual case) nasalised–words like *man* and *moon* are instances. Note, though, that in our model (in Part VI) these are instances of a cognitive *phonetic* rather than a phonological phenomenon.

phonology to feed to the disrupted area whilst maintaining its own accent-specific function. Put another way, model 1 is compatible with a practical extensible system while at the same time maintaining theoretical integrity.

8.2 Voicing

One of the features which has bothered researchers in the area of speech synthesis in the past has been voicing. We discuss this here because it is a good example of how failure to understand the differences between abstract and physical modelling can lead to disproportionate problems (Keating 1984). The difficulty has arisen because of the nonlinearity of the correlation between the cognitive phonological 'voicing' and how the feature is rendered phonetically. Phonological voicing is a distinctive feature–that is, it is a parameter of phonological segments the presence or absence of which is able to change one underlying segment into another. For example, the English alveolar stop /d/ is [+voice] (has voicing) and differs on this feature from the alveolar stop /t/ which is [−voice] (does not have voicing). Like all phonological distinctive features, the representation is binary, meaning in this case that [voice] is either present or absent in any one segment.

Voicing has been studied in terms of how the feature is rendered phonetically in terms of the physical parameters available. The focus has been both on the production of speech and its subsequent perception by a listener. One thing is clear from these studies: feelings about voicing match between speaker and listener. A listener's reporting of the presence or absence of voicing in any one segment in stretches of utterance matches the feelings of the speaker about the utterance plan. For example, speakers plan voicing in [b] and listeners perceive voicing in [b], and for lay listeners and speakers the results are consistent within any one language, but sometimes not between languages (Lisker and Abramson 1964).

The most frequent phonetic parameter to correlate with phonological voicing is vocal cord vibration–the vocal cords usually vibrate when the underlying plan is to produce a [+voice] sound, but usually do not when the underlying plan is to produce a [−voice] sound.

Many synthesis models assume constant voicing vocal-cord vibration, but it is quite clear that the binary distinction of vocal-cord vibration *vs* no vocal-cord vibration is not accurate. Vocal-cord vibration can begin abruptly (as when there is a glottal stop onset to make this possible–singers regularly do this), gradually (the usual case), or at some point during the phone, although it may be phonologically voiced. Similarly for phonologically voiceless segments, it is certainly not the case that on every occasion there is no vocal-cord vibration present at some point during the phone. Figure 10 gives some idea of the range of possibilities. We know of no model which sets out the conditions under which these variants occur.

These examples serve once again to underline a repeating theme in this book: phonological characterisations of segments should not be read as though they were phonetic, and sets of acoustical features should not be given one-to-one correlation with phonological features. More often than not the correlation is not linear nor, apparently, consistent–though it may yet turn out to be consistent in some respects. Phonology and phonetics cannot be linked simply by using phonological terms within the phonetic domain–such as the common transfer of the term 'voicing' between the two levels. Abstract voicing is very different from physical voicing, which is why we consistently use different terms for the two. The *basis* of the terminology is different for the two levels; and it is bad science to equate the two so directly.

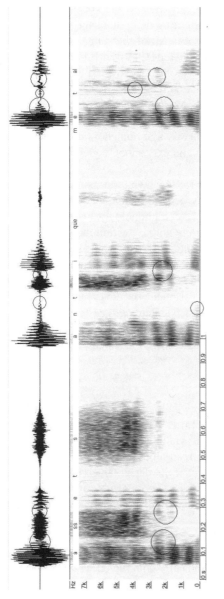

Figure 8.1 Waveform and spectrogram of three isolated words: *assets*, *antique* and *metal*. Areas of overlap or blend between segments are circled. Note how the frication of the [s] in *assets* overlaps with the vocal cord vibration associated with the adjacent vowels–events which cannot be modelled by any linear surface element sequencing model. In *antique*, the vocal cord vibration associated with [n] spreads into the stop phase of the adjacent [t]. In *metal*, again the vocal cord vibration of the previous segment spreads into the stop phase of [t], but this time the [t] itself is comparatively unusual in that it is laterally released; that is, the accumulated air pressure is released round the sides of the tongue before it is pulled completely away from the alveolar ridge. The appearance of the periodic source in the spectrogram at this point clearly reflects serious disturbances in the air flow, resulting in irregularities of vocal cord vibration.

For example, 'voice' *is not* 'periodic excitation'! Precisely the same error of science occurs in many automatic speech recognition systems when the normative phonetic specification for individual segments is given from the way they are classified in phonology. It is small wonder that misidentification follows so frequently.

As a rough guide around the contextual variations of the vocal-cord vibration parameter in synthesis, it is possible to return–pending a modern characterisation of all the observed variations in onset, offset, amplitude and spectrum of the source signal–to the early 'target + coarticulation' model (Hardcastle and Hewlett 1999) and derive some canonical form expressed perhaps in terms adapted from the abstract to the physical (phonology to phonetics) which alters with context. Coarticulation theory is often confused with respect to this context: sometimes it is expressed in terms of whole segments (and is almost phonological in the way it thus idealises the phonetic description) and sometimes it is articulatory or acoustic and parametric. The general theory is that somehow context degrades the ideal; but in terms of the science, a phonological context cannot degrade a physical object–only a physical context could do that.

The only rule which has proved regular for our own work is one which tells us that–still using voicing just as an example of a very general phenomenon–there cannot be a voiced phonetic object simply *because* there is a voiced phonological one, even if the meaning of 'voice' changes. The reason for this is that phonology, within the segment domain, is about constancy–there is *never* a change to the value of a parameter during the segment because the concept 'during' in respect of phonological segments is an error: there is no time in phonology. In sharp contrast there is almost always (perhaps *even* always) a change to the value of a phonetic parameter during a phonetic segment (in those models which admit of phonetic segments). It is also the case that these phonetic *changes* cannot be classified according to their phonological context–because the phonological context does not exist once the segment has been moved from the phonological level to the phonetic level. We believe it is not possible to overstate this point.

8.3 The Festival System

The Festival speech synthesis system (Taylor *et al.* 1998) is a text-to-speech system whose input is unmarked text and whose output is a synthesised speech waveform. It therefore aspires to the common objective of being able to read aloud unpredicted and unmarked text. The device is relatively transparent with respect to its design since one of the main objectives is to offer the system as a research tool. It is also extensible in the sense that users may augment the system by adding additional 'external' procedure sets or modules to achieve particular local requirements.

In common with almost all text-to-speech systems, Festival includes pre-processing of text to take care of abbreviations and non-orthographic symbols, a lookup system to enable phonological re-representation of orthographically represented words, a system for assigning a symbolic representation of prosody to the input sentences, and a final waveform synthesis arrangement, using concatenated waveform techniques.

The prosodic component of the system is like most other successful systems in that it involves a two-stage approach–the assignment of a symbolic representation which is later matched up with a corresponding acoustic rendering. The approach is not unlike that of Bell Labs' Multilingual Text-to-Speech System which is transparent enough in its approach to be transferable to different languages. In the first stage, phrasing and accenting partitioning is applied to phoneme-sized segment strings which are syllabically grouped, followed by a second stage in which fundamental frequency and other contours for the soundwave are computed from the symbolic markup.

The system is invaluable as a research tool provided the basic theoretical stance taken by the designers is adhered to. For example, researcher-users can substitute their own prosodic modules in the system–provided the principle of assigning symbolic markers to phonological representation and matching markers to the stored waveforms is kept. The modular design is important because it enables users to distinguish, for the most part, between high-level and low-level synthesis, keeping separate the different kinds of problem which arise in the different levels. A very important feature of the system is its theoretical integrity; and where this is not completely achieved, the designers are careful to bridge gaps expediently but also consistently.

It would be tempting to try to optimise different modules in a synthesis system by adopting different theoretical stances–this would create a system which lacked overall theoretical integrity, a sure recipe for disaster. Where choice exists, provided the theory is reasonable and accounts for a good percentage of the observed behaviour of speech, it would be better to choose integrity of conceptualisation rather than this or that mixed theory because each bit of theory is 'the best'. Designs like Festival and the much earlier Holmes text-to-speech system, or the Carlson and Granstrom (1976) design, have been successful for both research and in the field as practical systems precisely because they have the predictability of behaviour which stems from integrity within the overall model.

In the short term, basic systems represented by the Holmes or Klatt model (for parametric synthesis) or the Festival system (for concatenated waveform synthesis) are a good basis for taking forward ideas about naturalness. When adapted to niche domains they are excellent in terms of basic synthesis at the segmental level, and have all the potential for good prosody. They are certainly better than using pre-recorded prompts in, say, call centre applications. They are less successful, predictably, in large domains. It is enough to say, as a start, that what these systems lack is the ability to deal with variation at any level other than that directly predicted by phonological rules or phonetic coarticulatory rules. They fall short in having the means to handle more subtle types of variation, specifically expressive variation which is so essential to the perception of complete naturalness in the signal. To do this they will need to identify, at both the higher more abstract levels and at the physical levels, the parameters which carry this 'new' variability, and to provide the insertion points for taking input from new developments in the overall model, such as a well-defined pragmatics-sourced input. We expand on this in Chapters 36 and 37.

8.4 Syllable Duration

Phonetic syllables, rendered from timeless phonological syllables, have duration. Since the duration is unavailable in the higher level abstract representations on which phonetic rendering works, it has to be assigned during the rendering process. Duration is part of the rendering of utterance prosody, and as such has to be sensitive to pragmatic as well as plain message input; we discuss this elsewhere, and confine ourselves for the moment to the rendering duration without these extra nuances–though they will have to be there later if the resultant speech is to sound natural.

In computing syllable duration, most researchers have recourse to a hierarchical model which places the syllable in a higher level context. Syllable duration is dominated by such factors as utterance rhythm, as interpreted within the phrasal makeup of utterances. We might indicate the general model we use like this:

```
<rhythm>
        <phrasing>
                <syllable_duration/>
        </phrasing>
</rhythm>
```

Within the overall duration of the syllable will be the durations of the individual segments which the syllable node dominates. It is important to bear in mind that durational changes of the overall syllable do not correlate with the durations of its internal constituent segments. We know of no theoretically based algorithm for calculating the segment durations–though, of course, all systems need to do this calculation. Research within phonetics has observed that the correlations are nonlinear and lack stability, but we still await a consensus on precisely how to characterise the correlation (Tatham and Morton 2002). What is interesting is that phonetic theory allows, at the segmental level, for segments to have their own intrinsic duration which gets modified depending on the internal relationships of the syllable *and* factors external to the syllable, such as overall utterance rate and rhythmic structure. This can be viewed as a prosodic extension of the segment-based idea of coarticulation. The approach can be set out in a model as

rhythmic structure $\longrightarrow$ overall syllable duration $\longleftarrow$ segment durations

In practical synthesis, it is usually taken that it is necessary to calculate overall duration at the atomistic level of segment. Although this is at odds with how we believe listeners perceive duration (basically in terms of syllables rather than segments), the calculation works by starting with segment durations and modifying them in the light on how they work within the syllable and how they might be modified within the rhythmic structure of the segment. After this is done a further layer of modification is often calculated to take into account phenomena such as the slowing down of delivery rates toward the end of sentences and phrases within sentences.

A principled approach needs to ask how these various factors are linked in their common status as exponents of *prosody* in general. There are two reasons for insisting on this:

1 It is difficult to dissociate any single prosodic parameter from the others–and in fact may be a theoretical mistake to try to do so.
2 Additional external factors such as pragmatic considerations tend often to work on prosody as a whole, not just on individual parameters.

8.5 Changes of Approach in Speech Synthesis

The changes of approach in speech synthesis which we have seen over the past couple of decades have been mostly in the area of low-level synthesis: strategies for producing the actual waveform. There does not seem to have been a corresponding shift in high-level synthesis. Much of the phonological and prosodic requirements of text-to-speech, for example, are still as they were with previous systems. It is important not to confuse the high- and low-level components.

The main push forward has been in the refinement and general introduction of concatenative systems. Although strictly within the area of low-level synthesis, the use of a concatenated waveform approach–particularly where large units are involved–may well reduce the need for some areas of the high-level system, or indeed constrain its contents. We can see how concatenative systems model automatically much of the coarticulatory structure pronunciation–though not all of it, since cognitive intervention is not properly modelled in the usual concatenative approach. We should not confuse phonetic rules which account for some aspects of coarticulation and phonological which account for assimilation, but the results of some phonological processes will be included, of course in certain units. For example, in the case of word-sized units the phonological process in English which accounts for the use of two phonetically different [l] sounds will be included in any word beginning or ending with [l]. The only problem is that, with varying size units, or with small sub-word units, we cannot count on anything in particular being included. Certainly markup of the database and the phonological specification of the required utterance will have to be sufficiently detailed to prevent the apparent *double* application of rules; if something is already covered in the database it does not need to be computed before the units from the database are pulled into the main synthesis algorithm.

The use of a concatenative lower level does introduce problems of separation between phonology and phonetics in the underlying speech production model. In formant synthesis the problem does not exist; for, although most early text-to-speech parametric systems confused phonology and phonetics, today we can easily and productively separate the two. Assimilation and coarticulation need not be confused in such systems. But take the example of voice onset time associated with an initial phonologically voiceless plosive–say [t]. [t] before a stressed vowel usually causes the vowel to devoice significantly at its onset, except when the [t] is preceded by [s]. In addition, the degree of coarticulation varies with expressive content–with careful or stressed speech the coarticulation may be less–but this is not usually allowed for in concatenative systems. We know of no concatenated waveform system which succeeds in allowing for this variability either in a single database or between databases, though there have been proposals for multiple database systems (see Parts III and IX).

9

Automatic Speech Recognition

There has been enormous progress made in automatic speech recognition recently (Young 1999, 2000, 2002). But we might observe that successes in enabling larger vocabularies and more speaker-independent applications have been at the expense of being based on better models of human speech perception. In fact most of current automatic speech recognition is rather poor in this respect–which is disappointing to linguists or psychologists who tend to feel that automatic speech recognition which began as a promising simulation of speech perception has failed in this respect. However, the current model works rather well in terms of what speech technologists want it to do–recognise speech, not tell us how human beings perform the task.

There are three key properties in contemporary automatic speech recognition systems which have been responsible for most of the progress made during the last 25 years or so.

1 *Statistical modelling.* One of the main problems with automatic speech recognition has been the inappropriateness of linguistic and phonetic models. The idea of statistical modelling is to provide representations of speech objects based on the general properties distributed among them. That is, the objects are not just speech segments but also processes such as the transitions or coarticulatory phenomena which appear to link them. This is important because the general approach used is based on the 'sequenced object' model of speech which regards speech on the surface as a linear string of juxtaposed speech segments, with rules governing how they abut. This is the development of classical phonetics known as *coarticulation theory*.

2 *Learning.* If statistical techniques enable us to approach the modelling of surface variation in speech by introducing uncertainty into the equation, they can be really successful only if they are based on large-scale evaluations of data. Since researchers do not know the statistical properties of speech any more than linguists have modelled all the relevant details, it becomes necessary to scan and evaluate large quantities of data. This is impracticable for doing by hand, but is a task very suited to devices capable of data-driven learning. At present most automatic speech recognition devices use the data-driven statistical learning approach by developing, for example, hidden Markov-model (HMM) or artificial neural network (ANN) representations of speech during a learning or model-building phase. The resultant acquired statistical model is then used to process novel data to output hypotheses as to the linguistic content of the incoming speech signal.

3 *Hypothesis development.* Processing the waveforms to be recognised through the statistical models developed by learning results in a sequence of sets of hypotheses concerning the

Developments in Speech Synthesis   Mark Tatham and Katherine Morton
© 2005 John Wiley & Sons, Ltd.   ISBN: 0-470-85538-X

movement of the signal through the utterance(s). A contextually based evaluation system is usually able to narrow down the hypotheses to a plausible string of hits–solutions to the recognition task.

It is also possible to exploit various types of logic, so approaches which incorporate *threshold logic* or *fuzzy logic* are useful because again they help us avoid an approach which is based on certainties. The point is we need a means of representing both data and processes which are not fixed, but subject to variation. The problem with the abstract approach (even in early models like classical phonetics) is that it tends to minimise the variability found in speech. Variability is known to be very important; the concept appears repeatedly throughout this book–which is why it is necessary to understand its different types and sources.

9.1 Advantages of the Statistical Approach

Apart from retaining the idea of specially conjoined objects, a statistical model has little to do with phonetic or phonological characterisations of speech as known in linguistics. At best it constitutes a hypothesis concerning putative output from a knowledge-based model–linguistics is essentially a knowledge-based approach to characterising language. Such an approach has been shown not only to provide good answers in automatic speech recognition where the linguistics fails, but also to supply useful and important hypotheses for linguistics research in these areas.

One of the ways in which contemporary linguistics has proved inadequate in both automatic speech recognition and speech synthesis has been its static model approach, which is essentially about constancy or certainty. In speech, certainly at the phonetic level, static modelling is clearly only part of the solution to understanding what is happening; dynamic modelling of the kind we use here and elsewhere (Tatham and Morton 2002, 2003, 2004; Tatham 1995) is more appropriate because it offers techniques for incorporating

- time, and
- uncertainty (in the form of probability characterisations).

There was little or no attempt at dynamic modelling of speech in the early days of speech technology, and for automatic speech recognition the statistical approach offered the best solution for modelling both time and uncertainty.

The static modelling inherent in linguistics and peaking in Chomsky's 'transformational generative grammar' approach is very appropriate for capturing much of how language works in human beings. It is important to remember, however, that what is being modelled is the simultaneous underlying structure of the entire language. It cannot be overemphasised that linguistics, as approached by most contemporary theoretical linguists, does not come up with recipe models for creating this or that utterance. It characterises simultaneously *all* the objects and processes which, when interacting, establish the structure of *all* utterances–and the number of utterances, of course, is infinite. What researchers in speech technology systems need to know is all this *and* in addition how to tap into the static model to dynamically produce this or that utterance (speech synthesis) or recognise this or that utterance (automatic speech recognition). It would not be overstating the situation to say that most of what is needed is not in the characterisations provided by linguistics, because the dynamic processes involved in choice and selection which lead ultimately to instantiation are not included. This is not a problem in linguistics, but it is a problem in speech technology.

The success of the statistical approach to automatic speech recognition does not mean that the data-driven approach is inherently superior. It may simply mean that the earlier way of using rules and knowledge was inadequate because it was based on an understandable misconception of what linguistics is about. Running a linguistics description–a *grammar*–does not produce individual sentences (for semantics and syntax) or utterances (for phonologically and phonetically processed sentences); it produces what linguists call *exemplar derivations*. These are static examples of the kind of *idealised surface object* the grammar characterises; they are not particular examples of anything dynamically produced by the grammar. They certainly are not like the highly variable utterances people speak or the kinds of data automatic speech recognition devices have to deal with.

9.2 Disadvantages of the Statistical Approach

Despite demonstrable advantages there are distinct disadvantages to the statistical approach. One of these is the tendency for proponents to ignore the principle of plausibility in their solutions in terms of what it is that human beings do. This does not matter usually in the engineering environment, but it matters considerably in the environment where speech technology (both speech synthesis and automatic speech recognition) is used to test models of human behaviour in producing and perceiving speech–that is, where speech technology is a deliberate simulation of the corresponding human processes. But it can matter even if we are not really interested in finding out how human beings work. If speech technology is to be more than basically adequate in, say, dialogue systems it has to be sensitive to the *humanness* of speech. There is little doubt that modelling human speech production and perception as statistically driven processes is inappropriate. This does not mean to say that some parts of the dynamic processing involved in speaking or perceiving particular utterances do not involve some statistical processing; it means rather that the processing is not statistically based.

The statistical models do not gel with the linguistics (or psychology) models because linguists have taken the line which emphasises representations which capture maximal generality and which also focus on symbolic characterisations. They are right to do this because part of the way in which human beings clearly work is in terms of symbolic representations which tend to factor out variability. Linguistics and psychology tend to neglect modelling the relationship between these abstract representations which play down variability and the data itself which is exhibiting the variability. At the moment we cannot do much about this, but at least the nature of the problem can be explored and some tentative proposals for future directions made.

9.3 Unit Selection Synthesis Compared with Automatic Speech Recognition

The reason for opting for large databases in unit selection systems is to minimise the amount of modification and concatenation of retrieved units. When units are concatenated there is distortion. In most systems, two cost functions characterise the distortion:

- target cost–the estimated difference between the target unit (determined in the utterance plan) and the retrieved unit (determined by the database);
- concatenation cost–the estimated success of the concatenation of any two units.

Unit selection as a concept consists of searching and finding units in the database that minimise some combination of the two costs–usually their sum. The goal of the search is to optimise the acoustic rendering of the utterance plan. In speech recognition systems the goal is to hypothesise an utterance plan which best underlies or explains the input acoustic signal. Importantly, common to both systems is the utterance plan, which in our proposed model consists of a string of extrinsic allophonic segments wrapped in an equivalent prosodic plan. It is not possible to separate the segmental plan from the prosodic plan–symbolically they are integrated, as they are in the acoustic signal. The overall chain in the human situation is

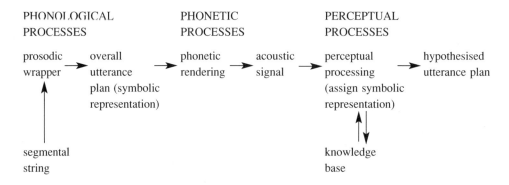

In the computer speech situation it is

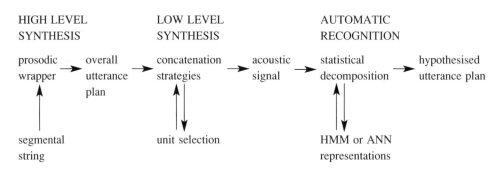

Both these diagrams omit expression and pragmatic references for the sake of clarity.

Part III

High-Level Control

10

The Need for High-Level Control

10.1 What is High-Level Control?

High-level control in our perspective on both speech synthesis and on human speech production is about processing taking place above the aerodynamic and acoustic systems. We think of speech production as comprising a basic two-tier hierarchy, with aerodynamic and acoustic systems occupying the lower level in the hierarchy. Although later we shall find it necessary to blur the separation a little, we find it helpful to use the two-tier approach in which the higher level is concerned in synthesis systems with *abstract processing*, and in human systems with *cognitive processing*.

> The necessity to blur a sharp division between the two components arises when we find that some of the normal low-level universal coarticulatory effects are able to be constrained at will. The theory associated with this we have called *cognitive phonetics* (though see also Lindblom's 'hyper-' and 'hypo-' articulation–Lindblom 1990, 1991), and rests on the notion of cognitive intervention in physical or biological processes in human beings.

The abstract/cognitive *vs* physical distinction is made as a matter of course in contemporary linguistics–phonology is that part of the model which accounts for cognitive processing leading, in production, to a plan for how to proceed to render an actual soundwave, and, in perception, to processing which assigns symbolic labels to the incoming acoustic signal. Thus in speech production we focus on the cognitive processes involved in formulating a speech plan for an utterance, and distinguish this from the low-level processes involved in rendering the plan. Rendering involves physical processes such as motor control, aerodynamics and acoustics. In speech perception, the low-level system involves acoustics and hearing, whereas the high-level system involves the cognitive processes involved in taking what is heard and giving it a symbolic or abstract representation (see Chapter 25 for a fuller explanation).

> In the human system, whether we are speaking of speech production or perception, the low-level system is primarily about physical phonetic processes and the high-level system is about cognitive phonological processes.

Developments in Speech Synthesis   Mark Tatham and Katherine Morton
© 2005 John Wiley & Sons, Ltd.   ISBN: 0-470-85538-X

In speech production the low-level processes include

- motor control
- the mechanics of articulation
- aerodynamic factors
- acoustic factors.

In speech perception the low-level processes include

- hearing–that is, acoustic analysis using mechanical means within the cochlea
- neural coding of the results of the analysis.

These low-level processes are physical. One question which might be asked is whether they are specific to speech. The answer is that they are probably universal in that they operate independently of speech–for this reason they are normally modelled within disciplines other than phonetics, or within phonetics using the terms etc. of some other discipline (e.g. neurophysiology, mechanics, aerodynamics, acoustics). This is important for linguistics/ phonetics because it enables us often to explain processes, events or phenomena having recourse to an *external* source–a valued way of proceeding for explanatory modelling in science.

The high-level processes are essentially cognitive. Within linguistics, cognitive processes involved in speech production and perception are treated within the phonological component, and to that extent are regarded as abstract (low-level processes contrast by being physical). In speech production, phonology characterises the processes and units involved in the sound patterning of language in general and in particular languages; in speech perception, the processes and units involved in assigning symbolic interpretations to an acoustic signal are dealt with by perceptual phonology. The treatments in linguistics and psychology are slightly different, with a different focus; psychology, for example, is often more concerned with perception rather than planning for production purposes, whereas the reverse is true in linguistics.

10.2 Generalisation in Linguistics

In recent linguistics, core components like syntax and phonology tend to be confined to static descriptive characterisations of their respective areas of language. However, if we move toward a more dynamic approach we can begin to characterise how individual utterances are planned.

It is important to note that contemporary linguistics focuses on generalisations. Indeed, generalisation and abstraction are central to the requirements of the theory. Little time is spent on specific utterance instantiations or on the variability which might arise during instantiation. If, however, we are to consult the theory of linguistics with a view to casting light on strategies that might be adopted in the design of speech synthesis systems, we shall have to make a dynamic approach more central. Generalisations are important, but, to invoke an extreme position, the abstract sound characterisation of *all* sentences in the language (which is what phonology, strictly, tells us about) may be less relevant than proceeding beyond this to a more dynamic description of what processes might be selected in formulating the plan for speaking a *particular* sentence. Thus static phonology is about the entire set of phonological processes which characterise the sound shape of *all* utterances within the language.

What is missing is a dynamic perspective which moves to characterising the processes involved in *selecting* or choosing from all these processes those necessary for a *single* instantiation, and how this single instantiation therefore differs from all others. The philosophy of the linguistics is thus in some sense turned on its head: generalisation loses no importance and must underlie any dynamic approach, but the *focus* of the study moves away from static general characterisation toward dynamic particular characterisation.

- In a *synthesis system*, it makes sense to us to draw the same line between abstract planning and physical rendering. One reason for this is that the two processes interface around the plan, in practical terms a file in which the basic requirements of the utterance are characterised. This characterisation has none of the variability associated with phenomena such as, for example, coarticulation. Thus the basic theory of speech production we use to support synthesis is the dynamic version of contemporary linguistics–the version which focuses on selection and choice from the general characterisation more common in contemporary linguistics. The abstract/physical distinction is made here as phonology (utterance planning) followed by phonetics (utterance rendering) is carried across to speech synthesis as *high-level* followed by *low-level* respectively.
- In *automatic speech recognition*, the principal task as we see it is somehow to derive the interface file (the perceptual equivalent of the plan), given the trigger of the input acoustic signal. The derivation, it turns out, is far from straightforward–hence the use of the term *trigger*. Certainly it is not a case of finding the plan *in* the acoustic signal or of the application of processes which equate the plan and the acoustic signal on a one-to-one or linear basis.

The interface file or plan is a representation in terms of extrinsic allophones (for the segmental material) fitted *within* a prosodic characterisation (at the equivalent of the extrinsic allophonic level). It contains all the necessary detail for rendering in production, or is the result of assignment in perception. The acoustic signal, in automatic speech recognition, constitutes the trigger for the processes which derive this representation.

It is difficult to show that the file is the 'same' for both production and perception; but it does seem to be the case (so far, we believe, unsubstantiated by formal experiment) that what a speaker feels he or she is producing (equating with the plan) matches well with what a perceiver feels he or she ahs heard. It would seem common sense that this is the case. There may be detail differences, but by and large we accept that this is at the very least a useful concept within the model: the equating of *production plan* and *perceived symbolic assignment*.

In speech synthesis, the plan file or its equivalent has another important function–it serves as the platform-independent input to the rendering process using *any* chosen synthesiser. The plan file looks the same whether it is to be rendered on a concatenated waveform system or a formant-based system (though some small set of adjustment rules may be necessary on some occasions). This is precisely in tune with the model of human speech production; the utterance plan is identical for any speaker of the same language or accent, although the way it actually gets rendered will differ between particular speakers due to their differing physical characteristics. In psychology, the utterance plan is also a useful concept because its identification by the speaker or listener is the basis of the human ability to recognise 'sameness' in differing renderings. When a listener says that the words spoken by Joe are the same as those spoken by Mary, the listener means that whatever actual sounds Joe or Mary happened to produce it is clear that the plan behind the two acoustic renderings was the same. A central task in speech recognition is to simulate this aspect of perceptual behaviour. We could use a term common in speech technology for this behaviour in human beings: the utterance plan, whether in production or perception, is *platform-independent*. Its character and composition are not a function of the platform–the human being who is going to render it or who has derived it. One of the major values to speech technology of linguistics is its ability to work with both platform-independent and platform-specific characterisations.

Notice that the idea of platform independence is well illustrated in systems using VoiceXML or SSML–two markup systems aiming at introducing a level of universality in speech synthesis (see Chapter 18). A principle underlying the specifications of these markup 'languages' is that all low-level systems be VoiceXML- and/or SSML-compliant to ensure transferability between systems with no further intervention. The idea we are expressing above is an *ideal*, and is rather more than the *practical* requirement needed for these two XML-based markup systems. We are taking the compatibility issue one stage further and giving it an underlying theoretical principle: human speakers are all different and produce different articulations and acoustic signals for similar utterances. They all are able, though, to work with an identical plan despite different rendering *strategies* or slightly different rendering *systems*.

Remembering the data structure (Tatham and Morton 2003)

```
<prosodic_plan>
      <segmental_plan/>
</prosodic_plan>
```

we have therefore in production and perception respectively:

Speech production

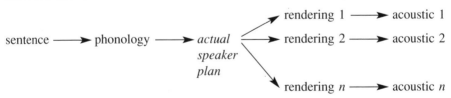

Speech perception

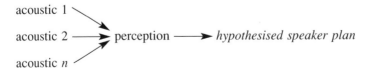

For us the *goal* of speech production is to produce a plan and render it such that a listener can derive a copy of it–the hypothesised speaker plan. The rendering and perceptual processes are therefore in a sense complementary, but at the same time are subservient to the plan itself. For this reason we feel that in both speech synthesis and automatic speech recognition the focus is appropriately the plan. In linguistic terms the plan is a string of extrinsic allophones (for the segmental aspect) rendered within a particular prosodic framework.

A useful example here is the ability to separate out assimilatory phonological processes from coarticulatory phonetic processes. Coarticulatory processes characterise the physically determined boundary constraints on abutting discrete speech segments. Thus those models of speech production which use phones or some similar speech unit at the phonetic or physical level need to account for the fact that at boundaries between these units there is not an abrupt change from one segment to the next–there appears to be some overlap or blend between adjacent segments.

10.3 Units in the Signal

The example of coarticulation just given underlines the difficulty speech researchers have in trying to identify objects like *phone* as the physical realisation of a phoneme. Phonemes, of course, are abstract units which can be thought of either as cognitive objects which underlie the plan of an utterance–a contemporary way of looking at what is happening; or they can be thought of as labels on groups of surface variants of the 'same' sound–a more traditional way of using the term. The problem is in trying to isolate phones in the soundwave. They cannot be delineated, of course, because they are coarticulated. It follows therefore that they cannot actually be 'found' in the soundwave–they are not there to *be* found. Phones are abstractions too, and we would prefer to call them *intrinsic allophones* (see Part VI). They are, in fact, symbolic representations of what is to be found in the soundwave.

This means that there is no point in saying that some object like a phone is a physical realisation of a phoneme; it is not, except in the comparatively rare situation where it might occur in isolation. In a normal utterance it is not possible to uniquely identify the phones or intrinsic allophones any more than the phonemes which might underlie them. The two most common units referred to in the synthesis and automatic speech recognition literature do seem to be phoneme and phone, yet, we repeat, it is necessary to underline that both are abstract (though one is more abstract than the other).

The problem is further compounded by the use sometimes of phone where we would perhaps use *extrinsic allophone* which is an abstract segmental unit in the phonological plan underlying phonetic rendering. We repeat that the nearest we can get to the term 'phone' in our own thinking is *intrinsic allophone*, which is a symbolic representation of a unit derived in the rendering process. An intrinsic allophone embodies something of its derivational history in the sense that it carries along with it how it has been derived. This is a common approach in linguistics (Wang and Fillmore 1961; Ladefoged 1965; Tatham 1971).

Models of speech which take allophones or phones as their basic units attempt to explain the observation of blending or overlap–the actual continuousness of speech–in terms of inertial effects at the aerodynamic, mechanical or motor levels in the rendering process. This explanation involves postulating that these constraints are not linguistic in origin, but constitute a true externally derived explanation of variability in the speech signal. As such these constraints have to be regarded as universal. That is, they are not language- or dialect-specific, but are there because of the physical design of human beings themselves.

But not all blends or overlaps fall within the scope of coarticulation theory. There are overlap variations which appear to be optional or voluntarily determined. An example of this is the assimilation of an intervocalic /t/ which is ordinarily a phonological [−voice] stop to a [+voice] flap. This occurs in some accents of English in a word like *later*–often where the first syllable is stressed and the second unstressed. Although we might predict the direction of the assimilation (that the [−voice] element should become a [+voice] element and that its [+stop] feature should change to a [+flap] feature, we cannot predict its general occurrence because it is not obligatory. For this reason we do not call this phenomenon coarticulation but rather assimilation, making it fall squarely within phonology (within the domain of cognitively sourced choice) rather than phonetics (the domain of involuntary physical constraint)

We shall see in Part VI, however, that the universality of coarticulation is not entirely safe as a concept because there are occurrences even here at the physical level of cognitively driven options. So, for example, although the nasalisation of a vowel nucleus in syllables like [. . . mun . . .] and [. . . mæn . . .] occurs at the phonetic or physical low-level, it is optional (therefore cognitively driven) just how much nasalisation there will be. There cannot be *none*, however–which is why the process belongs in phonetics. In instances like this we see a clear overlap between physical and cognitive processes which are often systematic–a consistently high-level of nasalisation is found in some American English accents of English compared with others, or compared with many British English accents.

10.4 Achievements of a Separate High-Level Control

The introduction of explicit high-level control improves *general* versatility by enabling the input of

- plain text
- annotated text (of different types)
- 'concept' etc.

And it improves *specific* versatility in terms of enabling detailed control of specific types of input, such as giving us more direct control over local variations like the prominence of particular words within a sentence.

10.5 Advantages of Identifying High-Level Control

Identifying high-level control enables us to separate out cognitive from physical processes in the model of human speech production, and in speech synthesis enables us to identify processes which are not platform-dependent. In speech synthesis there is the additional benefit of enabling us to work with a model which is sufficiently like our model of the human being to justify the use of the term 'simulation'.

So, using the notion of high-level control in speech synthesis enables us to match up with an explicit model of human speech production, inheriting its advantages. This transparency of modelling strategies between human and synthetic also enables the synthesis system to function as a testbed for the human model–an important use of synthesis. Since in the human model the high level is where we model those phonological processes which are also cognitive, we can make sure that we are as compatible as possible with approaches taken in psychology–thus explicitly extending the scope of the testbed function of speech synthesis to modelling some psychological processes as well. For a number of reasons (e.g. the strict scientific basis of the approach to speech planning in linguistics may not be identical with the approach in psychology) compatibility cannot be guaranteed here, but awareness of the source of the data and the general modelling approach is certainly better than any unprincipled matching of models with potential test environments. A simplistic gain would be the highlighting of areas of differences in detail between the linguistics and psychology approaches. The advantages of working to unify psychology and linguistics are more apparent in attempting to match up automatic speech recognition strategies for assignment of symbolic representations with those adopted in linguistics and psychology; the reason is that psychology tends to focus more on perception when it comes to speech modelling.

Once we have identified high-level control we are able systematically to deal with many of the features found in text-to-speech systems. In particular we are able to accept inputs from a number of different sources–for example, a plain text input channel supplemented by a 'mode of expression' input channel which might originate in some pragmatic processing arrangement to give the final signal expressive content.

11

The Input to High-Level Control

We are able to identify three principal input sources for high-level control in our model of speech production:

- linguistic (segmental)
- linguistic (prosodic):
  - intonation
  - stress (focus, prominence)
  - utterance rate
  - rhythm
- pragmatic/expressive (prosodic).

11.1 Segmental Linguistic Input

The underlying sentences to be spoken are derived in areas of language which are little connected with speech. On the surface, sentences are ordered strings of words, and are an encoded representation of ideas or thoughts which a human being wishes to communicate to one or more other human beings. The ordering follows the syntactic grammar characterised within the syntax domain on the model's static plane (see Chapter 22). Any one sentence for speaking has been derived on the dynamic plane using this grammar; any one sentence derived on the static plane is an exemplar derivation, not one intended to be spoken using the normal dynamic phonological and phonetic channels.

The segmental linguistic input to the speech synthesis model is the least controversial of all the areas within synthesis simply because it is given. It is the text input to a text-to-speech system, or it is the sentence encoded output from a concept-to-speech (Young and Fallside 1979) system. In text-to-speech systems which are designed to read text out loud, the detail of the text may have different properties which are more or less explicitly defined depending on the detail included. So, for example, the text may have been

- written to be spoken, like a drama script including stage directions and the other conventions associated with scripts
- designed only for silent reading, like a book such as this one

---

Developments in Speech Synthesis   Mark Tatham and Katherine Morton
© 2005 John Wiley & Sons, Ltd.   ISBN: 0-470-85538-X

- abbreviated using a small set of simple conventions, like early telegrams
- abbreviated using complex socially motivated but changing conventions, like contemporary texting.

We look elsewhere (see Part V) at how most text-to-speech systems augment text before it can be phonogically interpreted. This is a process which expands abbreviations and other symbolic conventions like figures, and in modern systems integrated into, say, VoiceXML applications (see Chapter 17) adds either manually or automatically various levels of markup to make explicit objects which in the original text are phonologically opaque. Examples of opaque text objects are: 'Dr.' (*Doctor*, or *Drive*), '*et al.*' (*and others*), '546' (*five hundred forty-six*, or *five hundred and forty-six*, '88.93' (*eighty-eight point nine three*, or *eighty-eight point ninety-three*), '13 deg. C' (*thirteen degrees Celsius*, or *thirteen degrees Centigrade*), 'the year MMV' (*the year two thousand and five*, or *the year two thousand five,* or *the year twenty* [ou] *five*), 'C 1602 W. Shakespeare' (*copyright sixteen* [ou] *two by double* [ju] *Shakespeare*), 'I luv NY' (*I love New York*), etc. (cf. Allen *et al.* 1987).

11.2 The Underlying Linguistics Model

The linguistic segmental input is a symbolic representation of the plain message content of what is to be communicated. In traditional linguistics terminology the domain is the plain sentence. It is characterised by exhibiting no optionality or variability–the sentence has been formulated as an appropriate abstract representation of what is to be communicated in terms of the available words ordered in an optimum way. It is important to note that a sentence is a highly abstract object–for the moment it even has no existence outside the cognitive capability of the speaker. For any one act of speech this sentence represents a specific instantiation, a particular sentence from the infinite set of all sentences in the language–the one which best does the job in hand. For the moment it is completely unpronounceable.

Segments for pronunciation usually take the form of units which are said to be 'phoneme-sized'. An alternative is syllable-sized segments. Syllables are longer than individual segments and usually comprise a string of segments obeying strict hierarchically based rules. Researchers are not completely agreed as to what shall be the basic segments for modelling speech. Phonemes, for example, are very highly abstract units well-removed from the acoustic signal by at least two levels of rule sets: phonological and phonetic. They are the ultimate high-level unit. It follows that labelling an acoustic database with phoneme symbols makes a nonsense of attempts at alignment: to align a representation this abstract with the actual acoustic signal is not a sound practice scientifically. Labelling is *per se* not a problem; it is the temporal alignment which is the problem because abstract units cannot have temporal boundaries in the way that physical units do.

'Phones' are the units preferred by many researchers. These too are abstract, but not as abstract as phonemes. Phones are said to embody all phonological process–they are thus units at a sub-phonological level. The question is whether they embody all or some phonetic processes as well. So, for example, what would be the label used for the underlying phoneme /d/ in a word like *width?* The sound has very little or no vocal cord vibration associated with it, and moreover has dental rather than alveolar constriction. If we incorporate both these features into the phone representation, then we probably have the lowest level practical linguistic representation–a representation of both selected phonological and phonetic features, though several types of variability will not be included even at this level (e.g. expressive variability).

Because the definition of phone is so vague in the literature we could revert to choosing between two terms from a more recent theory which attempts to make sense of *allophones* and their distribution. In classical phonetics, allophones are variants of phones derivable from a single underlying phoneme. The point is that there *are* clear definitions for the terms *extrinsic* allophone and *intrinsic* allophone. We repeat the definitions here, although we use them elsewhere, because they are so fundamental to the model.

- An *extrinsic allophone* is a symbolic representation of a speech segment which is used in the characterisation of a speaker's underlying plan for an utterance. As such is it abstract and representative of cognitive *intention*. The variability reflected in extrinsic allophones arises in phonology. A good, common example is the planned systematic usage of velarised and palatalised /l/'s in English; the usage of one or the other /l/ is entirely voluntary and can easily be changed at will.
- *Intrinsic allophones*, on the other hand, reflect *in addition* a variability which is not determined cognitively. A useful example here is the dentalised [d] in the word *width* mentioned above. Here an alveolar constriction is at best difficult and at worst impossible to make, depending on the rate of utterance.

Extrinsic allophones are symbolic representations of segments which have been cognitively derived, within their syllabic context, from the extreme abstract underlying phoneme. The variability they capture is phonological. Intrinsic allophones, on the other hand, incorporate a *further* level of processing (not an alternative level of processing), one which is not cognitively sourced. This is the level of processing characterised in coarticulation theory–processing which results from mechanical, aerodynamic and other constraints on articulation. The variability captured in the intrinsic allophone is both phonological (earlier derived in the phonology) *and* phonetic (derived during phonetic rendering of the corresponding extrinsic allophone).

Markup of the acoustic signal in the database associated with a concatenated waveform system may well need to be multi-layered even in terms of segments. There is good reason for wanting to use extrinsic allophones–they enable markup directly relevant to the *utterance plan*, since they are the units it is expressed in. But there is also a need to capture the immediate phonetic context, and this can be expressed with an intrinsic allophonic markup; this will capture a characterisation (abstract) of coarticulation as applied to the extrinsic allophones (not the phonemes).

We need to think very carefully about the level(s) of representation for useful markup of a database; it is by no means obvious how this should be done. Even at just the segmental level, multiple markup may be necessary. Perhaps some levels of markup can be *derived*. For example, if we understood enough about coarticulation then extrinsic allophone markup might well be enough since we would be able to derive an intrinsic allophonic markup directly. The problem with a markup which reflects context is that the influence of context is itself a variable which is not obvious from the segments themselves. The reason for this is that it is not just the adjacent segments which determine the coarticulatory context but the segments in conjunction with temporal variation. To simply mark up a text to show general contextual influence may not be enough. To repeat: ostensible coarticulation depends on segmental context, but in fact the extent to which it depends on adjacent segments varies with the continuously varying timing of these segments. The context of coarticulation, usually modelled as a static phenomenon, is in fact a dynamic phenomenon. We shall see later that exactly the same is true of prosodic markup–the varying context means that the *markup itself* has to be variable and is by no means obvious.

11.3 Prosody

Phonological prosody refers to the non-segmental phenomena of intonation, rhythm and phrase or sentence stress. These are responsible for encoding into speech some syntactic and semantic information about the underlying sentence. Prosody is also used as a vehicle for expressive content–encoded information which is modelled as distinct from the more basic semantic and syntactic information. Expressive content is mostly sourced from a pragmatic component in linguistics.

The abstract phonological parameters of prosody have physical exponents: fundamental frequency, amplitude and rate-of-delivery *contours*. Contour is the term used to characterise how these three parameters change over time: it is the *way they change* which encodes the prosody. When rendered the way these physical exponents behave is called *phonetic* prosody.

The greatest difficulty we have in modelling prosody is the fact that the relationship between the parameters of phonological and phonetic prosody is not only nonlinear but the way in which the physical exponents combine to reflect the three underlying abstract parameters itself varies. To repeat: we have three underlying abstract parameters–intonation, rhythm and stress. Each of these is encoded in varying combinations of three physical parameters in a nonlinear and badly understood way. To make matters worse, the way in which a putative 'basic' or expression-free prosody is overlaid with expressive content is not understood either, with the result that the final acoustic signal is, for the moment, very difficult to model transparently.

When sentences are to be spoken they need converting into phonological representations; this process is the abstract planning of particular acts of speech–utterances. These phonological representations involve a complex structure of syllables and their component segments. However, the structure of the sentence itself (its syntax), and a number of other factors, influence the way a phonological representation takes shape. From our perspective the phonological representation is built within a prosodic framework which needs to be determined for the utterance; this constitutes the linguistic–prosodic input to the high-level speech production or, in our simulation application, to the speech synthesis system. By beginning not with how the segmental characterisation of an utterance is built, but with its prosodic structure, we have introduced the idea of a prosodic *wrapper* for speech, and we shall continue this idea here. In modern times this fundamentally different approach from the notion that prosody is fitted to a characterisation of the segmental structure of the utterance comes from early proposals by Firth (1948), among others.

The linguistic prosodic input to rendering itself constitutes the prosodic plan. If we introduce the same kind of approach as with the segmental input, we can say that the prosodic structure of an utterance rests on some idealised structure associated with the particular sentence to be spoken. This is parallel with the extrinsic allophone idea. Prior to this, though, there might be a phoneme-like equivalent. So we can imagine a static phonology whose job is not just to enumerate the sound shapes of all possible utterances, but to enumerate them within all possible prosodic wrappers. For any one individual utterance a parallel dynamic process builds a single utterance plan by drawing on the possibilities characterised on the static 'plane' of the system. The utterance plan thus constructed is wrapped in an appropriate prosodic structure. The simplest possible justification for looking at things this way round is that there are clearly segmental properties of an utterance which are prosodically constrained, and to obtain these in the traditional approach the prosody-free utterance plan would have to be rewritten in terms of a prosody fitted to it. The two approaches are illustrated here:

Prosodic wraps segmental

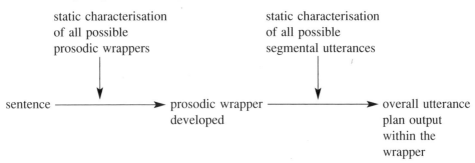

Segmental wraps prosodic

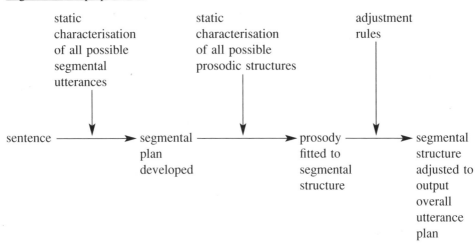

What would the characterisation of prosodics look like? For segments there is language-specific selection from among a universal set of all possible phonemes. If this is the case then there would be a similar language-specific selection from among a set of universal prosodic contours, which will be multi-parametric: intonational, durational, rhythmic. For this language and this sentence there would be a derived sequenced set of such contours–equivalent to the extrinsic allophonic representation. Indeed the extrinsic allophonic representation is wrapped by this 'extrinsic allophonic prosodic contour'.

11.4 Expression

We have extended the idea of wrapper to include a pragmatic/expressive input to the high-level speech processing system. We see the linguistic segmental and the linguistic prosodic inputs themselves formulated within an all-embracing expression wrapper. All human speech has expressive content and it is remarkable how little attention has been paid to this in earlier models (Tatham and Morton 2004). Although contemporary practice has shown that it is possible to model speech of some kind without any consideration of expression, the result (on the human front) is a prediction that speech is expression-free (which is false), and on the speech synthesis front an acoustic output which is expression-free (which is unnatural, and sounds it).

The pragmatic/expressive input for an utterance wraps the whole utterance plan in its expressive environment. The plan representation here is still abstract. As yet the constraints on rendering have not yet operated–these come in when rendering actually takes place on the dynamic phonetic plane of the model.

It is worth a brief note at this point to mention that there are constraints on speech introduced because of the nature of the vocal system. These constraints are not only imposed on what the theory sees convenient to characterise as individual segments, but also on the prosodic units. These constraints operate globally, and ultimately determine what is and what is not possible segmentally and prosodically for use in spoken language. The constraints influence not just phonetic rendering but also what is available for cognitive processing–languages are not going to engage in phonological processing of objects which cannot be rendered. This set of universal underlying constraints is different from the set which kicks in at run time (i.e. during *actual* rendering), though the parameters of the constraints may appear to be identical.

So there are

- *global constraints* on rendering which are responsible for the universal inventory of human speech sound segments and prosody, and
- runtime *constraints* which influence how what *is* possible is rendered.

For a presentation of expressive context, and how it is incorporated into the static/dynamic model, see Chapter 22 and Chapter 24.

12

Problems for Automatic Text Markup

A major problem for automatic text annotation is how to extend the annotation beyond a simple phonological segmental markup. Such markup has to be augmented with general prosodic and expressive prosodic information which are not encoded, except perhaps in a very rudimentary way, in plain text. If text-to-speech is a simulation of the human capability of reading text out loud, then it is important to understand that during this process prosody and expression are *not* extracted from the text by a human reader–they are assigned from within the reader's cognitive processing, and used as a *framework* within which to deal with the text input. If a speech synthesis system is to fully simulate a human speaker than it has to deal with these factors too; a plain text input will not be sufficient to produce other than a rough expressionless and prosodically unlikely version of the text. Thus, just as in the human reader, the synthesis system must actively assign to the signal the missing prosodic and expressive information.

With concatenated waveform synthesis, especially when incorporated within the unit selection model, the markup of the database is critical to enable proper identification of the optimum stretch of waveform for selection; a mistake here, particularly in alignment of the markup, can cause serious errors in the final output.

Unit selection systems will fail if they make predictions concerning this or that unit for inclusion in the output based on errors in markup. Markup procedures need therefore to be sufficiently robust to minimise selection errors. The very minimum is an appropriate and accurate identification of phone and sub-phone sized units, together with a way of indicating the prosodic wrapper they were produced in for the original database recording. However, hand markup is very time-consuming, and various researchers have sought to introduce fully or semi-automatic markup systems.

We stress that the wrapper itself needs to be marked. The reason for this is that there are global and local prosodic effects which cannot be specified by just marking the short-term measured values for fundamental frequency, amplitude and duration. Prosodic effects, both basic and encoded expressive content, rely on contour–meaningful changes in value expressed as trends in direction. The content is encoded in the *contour*, not in any instantaneous value. There is a parallel in segmental material: formant transitions between segments encode information about the segments themselves, but no reading of an instantaneous value for frequencies or amplitudes of formants will reveal this information–it is the trend or contour of the formant values which holds the encoding.

Developments in Speech Synthesis   Mark Tatham and Katherine Morton
© 2005 John Wiley & Sons, Ltd.   ISBN: 0-470-85538-X

For both segmental and prosodic information we need markers which indicate phonological/phonetic trends running between a certain number of phones to the left and right on the current phone. This information is in terms of

- adjacent segments (symbolic segmental context)
- fundamental frequency trends (relatively short term for local prosodic effects like emphasis or focus, and longer term for sentence and paragraph intonation)
- amplitude values (short term for word stress effects and long term for contrastive and prominence effects)
- duration values and trends (short term for rhythmic units, long term for utterance rate).

12.1 The Markup and the Data

Dividing the synthesis production model into two basic levels, high and low, leads us to ask where we have been getting our data from for these two levels, and importantly whether the data sources are compatible when marked up.

- Generally the source of data for the low part of synthesis has been a survey of the acoustic characteristics of one or more speakers' voices. This is expressed in formant systems as a general parametric model of the target waveforms for various phones. In concatenated waveform systems it is usually confined to the speech of a single person per database. In these systems, depending on the size of the database, the acoustics of the speaker is more or less exhaustively modelled by an actual waveform collection designed to capture as many as possible of the properties of the speaker's speech.
- At the high level the model has been in terms of linguistic generalisations about one accent of the language in question, usually without reference to any particular speaker. There is a basic assumption here that the high-level part (the phonology) of any one accent is not speaker-specific, and in general this assumption is reasonable. High-level processing is distinct from the database and involves procedures applied *to* the database, but high-level *data structures* are usually marked up directly on to the acoustic signal.

Our feeling is that provided the division is made rigorously the sources of the data are probably compatible. Care would obviously be taken to make sure that the phonology which has been captured indeed expresses what underlies the speech of the particular individual used to record the database of a concatenated waveform system. Similarly in a parametric synthesis system, the model will be either of a single speaker, or some kind of average of a number of speakers. Again care will be taken to make sure the phonetics (the low-level characterisation) and the phonology (the high-level characterisation) match.

One problem which does arise, though, when it comes to parametric systems is that in modelling some kind of average speech there is the risk of modelling no speaker at all. This apparent paradox has two sources:

1 Statistically there may be no actual scores in the original data which match any one average score. It is easy to illustrate this by taking two scores, say 9 and 11, and averaging them to 10–there is no actual measured piece of data with a score of 10. The average is therefore an abstraction from the data. Everyone who has considered statistical

means of data reduction knows this; and they also know that we need data reduction in speech because of the quite considerable spread of the data in terms of its inherent variability.

2 What is unfortunate, though, is that this very variability is itself responsible for much of what listeners use to judge speech they hear to be human rather than synthetic–or good rather than bad synthesis. In averaging data to produce a kind of idealised specification, the variability itself will almost certainly be lost. Similarly there is a danger with concatenated waveform synthesis that the database may not be large enough to capture enough natural variability. If properly constructed it will be large enough to capture all co-occurrences of phones as described by phonetics and phonology; but these, for one reason or another, say little about the nature of the variations which also need to be captured. It is likely that to do justice to natural variability in speech the database would need to be vast to capture not only the usual co-occurrence but many examples of *each* co-occurrence. The extreme case here was found in early diphone systems where, in the tradition of early formant synthesis with its single target for each phone, the database consisted of a single diphone for each pairing of phones.

In short: variability goes way beyond the idea of constraints due to segmental context, whether this is phonological or phonetic. Fail to capture this extra variability and listeners reject a system as producing unnatural speech.

Another way of saying this is that in concatenated waveform systems there are several parts. One is the explicit modelling of the acoustic database–simply by labelling it. This incorporates as much variability detail as the markup philosophy calls for or permits, since it must have some constraints and cannot be open-ended. But the markup system itself (or at least the linguistic part) is couched in terms of generality–it is not about *this* speaker's linguistics but about the linguistics of the language; any individual speaker variability is normalised or abstracted out, simply by neglecting it. Although we have said that this is probably reasonable, when considering that the errors arising from neglecting *phonetic* variability are probably much greater, this will inevitably eventually lead to mismatches between the markup and the signal.

12.2 Generality on the Static Plane

Linguists and psychologists take the view that it is important to focus on models which capture maximal generality and which also focus on symbolic characterisations. The two things are not necessarily linked since, obviously, there can be generality among physical phenomena without recourse to symbols. Linguists are right to do this because part of the way in which human beings clearly work actually is in terms of symbolic representations which tend to factor out variability. Linguistics and psychology tend to neglect the modelling of the relationship between these abstract representations which play down variability and the data itself which is exhibiting the variability.

It is important for speech technologists to understand the reason for this. For both speakers and listeners when they are *considering* speech (i.e. when they are thinking about it or processing it cognitively), the variability which we find in waveforms in the physical world is for the most part not known. That is, speakers and listeners are not explicitly aware of waveform variability; if the variability is noted it is detected only at a subconscious level.

But, remarkably, remove the variability and speakers and listeners alike will be immediately aware that 'something is wrong', though they may not be able to say what this is.

The paradox is that linguists and psychologists–essentially concerned with the cognitive aspects of language–generalise out this a-linguistic variability since it does not contribute to the meaning as encoded in the signal or assigned to the signal by a listener. Meaning here means meaning of the intended message. But there is another meaning. Part of the meaning of speech is that it is human, and this meaning is conveyed very much by inherent properties in speech (like its variability, or the fact that there are formants, or the nature of the periodic source waveform etc.)–they express the humanness of speech though not the meaning of any one message.

Automatic speech recognition attempts to reduce the variability and derive symbolic representations based on a speech wave input. This is what human listeners do, though perhaps not in the same way as most automatic speech recognition systems. But we are concerned with synthesis, and what we have to do is introduce as much human variability as possible into the signal–whether it represents the message or not, so long as it means that this is a waveform which could have been spoken by a human being. Naturalness in synthesis *means* creating a signal which listeners judge to be consistent with what a person would produce.

The generalisations of linguistics, to be found on the static plane of our own model, characterise in principle the entire language–or at least, if the characterisation is incomplete, part of the language–in the most general terms possible. Almost all synthesis systems which have adopted the descriptions of linguistics make the serious error of deriving characterisations of individual utterances by direct application of the linguistics rules. What this does is produce an *exemplar sentence* (for syntax and above) or an *exemplar utterance* (when phonology and phonetics are brought in). An exemplar utterance is one of the entire set of idealised utterances which comprise the language; it is *not*, formally, an instantiation. An instantiation for us is an utterance as produced dynamically by a human being, in part by applying their (static) knowledge of the language. Formally it is simply a coincidence–and an unfortunately confusing one–when an instantiated utterance appears identical to an exemplar utterance. The two are not equivalent. At the very best an exemplar utterance will sound strange (because it lacks essential variability not characterised in linguistics), and at worst will be quite wrong because it fails to capture vital expressive content (see Chapter 23 for a fuller explanation).

12.3 Variability in the Database–or Not

However, if we are going to continue to synthesise speech from units found in large databases, we could consider how we are going to record such databases and then how we are going to mark them with all the additional information needed for more realistic synthesis. The further information annotates the extra material recorded with more segmental and more expressive variability. An alternative approach would be to use the database in much the same way as it is now–mostly for just the segmental aspect of speech–and generate the remainder on demand. This would have to be done with any synthesis technique (concatenated waveform or parametric: unit or abstract target extrinsic allophone)–so the systems start to reconverge and the differences between them become less obvious as the focus moves toward more satisfactory coverage of expression. Neither of them does this well without the addition of something not contained within the original inventory–they both have inventories or databases, though of differing types.

Multiple database underlies varying output

Plain database transformed using a variability model to produce varying output

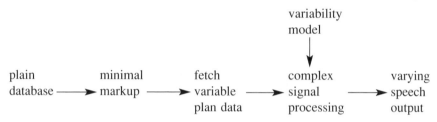

Whether the database intended for advanced synthesis involving expression has multiple sub-components or not, it is still the case that the markup or processing needs to be extensive and sophisticated. Switchable or multiple databases are designed to allow several voices or several different types of expression all within the same basic system. In principle the idea should work, but it is of course fundamentally unsound from a theoretical viewpoint. It is highly unlikely that people switch underlying databases when they switch expressive content, and indeed in the case of biologically sourced emotional content it *cannot* be true (see Chapter 23).

The idea of having multiple databases is to get around the fact that we lack a good model of variability in speech production–that is, specifically the variability due to express-ive content. By having switchable databases (say, one for persuasive speech and one for authoritative speech) it is hoped that we avoid the need to model the acoustic properties of these and other types of expression. The alternative is to have a single database devoid of expression and modify it on demand to convey expressive content.

The drawbacks to the multiple switchable databases approach are several:

- There are practical difficulties associated with making databases using a single speaker (for different expressive content) or different speakers (for different voices) with con-sistent recording characteristics–the problem of exponentially increasing size.
- A comprehensive and necessarily complex markup system is needed to indicate all usable content, both segmental and prosodic.
- A complex system is required for interpreting the markup and conditioning it ready for procedural handling to create the appropriate waveform modifications.
- It cannot perform short-term changes, say of emotion, or progressive blending of express-ive content as the speaker changes attitude *during* an utterance (a problem with the exclusivity of the different databases or sub-databases).

The drawbacks to the plain database approach are:

- There is a lack of suitable models showing the relationships between different.voices or the acoustic characteristics of expression.
- It provides poor signal processing techniques for handling frequency domain changes (including modifying individual formant characteristics).

The advantages of the multiple switchable database approach are:

- There is absolute consistency, if that is what is needed. Long-term characteristics are maintained properly.
- A well-designed markup system can characterise considerable subtlety in the prosodics data structure.
- There is explicit comparison and rendering of utterance plans.
- There is comparatively simple signal processing.

The advantages of the plain database approach are:

- The data collection effort can concentrate on large-scale recording without the need to use 'space' for multiple voices or expressive content. There is room for improvement in naturalness due to multiple choices between units.
- Expressive modelling is more in line with what human speakers do.
- Much of the signal processing, though complex, falls into line in explicit accordance with the underlying model of expression–probably improving naturalness.

In practice there may, in the end, be little to choose between the two approaches, though of course the choice is more sound if it is well informed. In addition, choice may well depend on application and availability of resources. There is clearly a trade-off between advantages and disadvantages between the two methods.

Whichever approach is chosen, we need to be clear about isolating what may be called in the abstract our *information channels* (sources of constraint on how to formulate the utterance plan and render it) from the *physical vehicles* needed for satisfactory rendering of the final acoustic. Whether expression wraps prosody which in turn wraps the segmental content of utterances (our clearly preferred option), or whether segmental plans acquire added plain prosody which is then transformed in the final acoustic signal to convey expression, we shall still need to identify:

1 *information channels*

- segmental stream
- generalised prosodic contour
- intended style of delivery
- speaker-explicit or inexplicit expressive content

2 *dynamic physical vehicles*

- timing and rhythm control
- on-demand modification of the voice source characteristics
- fundamental frequency control for contour
- modification of spectral detail.

Information channels and dynamic physical vehicles need to be isolated and fully explicit because of the nonlinear relationship between them, both vertically (the relationship between information and rendering vehicle) and horizontally (how the different channels play off against each other, and how the different vehicles interplay for rendering). Unfortunately there are no final answers yet about such things, but at least we are now in a position to enumerate them for developing hypotheses and conducting experimental investigations.

12.4 Multiple Databases and Perception

We have seen that the inclusion in speech synthesis systems of different databases from speaker(s) exhibiting different emotions or expressive content is one proposal for developing concatenative natural synthesis. We have also seen that there are other ways of introducing the expression–for example, by manipulating by rule and on demand the acoustic signal from a single relatively neutral database.

We are making certain demands on the listener, and the resolution of these demands constitutes in a very real way the *goal* of what we are trying to do–deliver a signal to a listener so that he or she perceives everything correctly, and in such a way that the listener can report having listened to a real human being. We are presenting to the listener a physical encoding of *something* in *some way*. For the moment these are vague terms to focus our attention, as scientists and engineers, on the independence of the plain message and how it is spoken.

The task of the listener is to *assign* to the signal an *interpretation* leading to being able to report what the message is and something about the speaker (attitude, feelings, and so on). We are careful to use words like 'assign' and 'interpretation' because it seems to be generally agreed (though there are some researchers with strong objections) that the final reporting by the listener is a kind of mix between what was in the signal and what the listener has brought to the signal in an active way. Another way of saying this is that the task of the listener is to use the acoustic signal as a trigger for a complex *active* process of interpretation which draws on a great deal of what the listener knows about the language, speech, the speaker, the environment etc. Perception is not a passive experience which extracts information from the signal: *it is not there to extract.*

The robust link here is an agreement between speaker and listener as to what has been encoded, but no real conscious knowledge on the part of either as to how this has been physically achieved either on this occasion or on any other, or indeed how anything is encoded into utterances. On the face of it this appears to be a considerable paradox, but it rests quite simply on what Chomsky (1957) called the 'tacit knowledge' of the speaker/ hearer. Chomsky was talking only about knowledge of language and how it works, but we can extend the idea to all or most speech perception: as speakers and listeners we really have no conscious or reportable idea what we are doing or how we do it.

12.5 Selecting Within a Marked Database

Markup of the database in a concatenated waveform system is critical because it is the only means into selecting appropriate units from the data. The waveform itself is not used directly in the selection process. It follows that the selection process relies heavily on

- the *amount of detail* in the markup
- markup *accuracy*
- the *sophistication* of the decision-making process
- the *relevance* of the criteria used in the process.

The procedure underlying an act of selection depends on criteria deemed relevant and how they are applied in the selection algorithm. The goal is to produce a sequence of labels which can be searched in the labels annotating the database. Identity of plan labels (for a particular sequence of units) and database labels means that the plan and waveform are the right match. Any two identical sequences of markup labels on the database means that these sections of the database are themselves identical–identity is determined *not* by the nature of the signal but by the composition of the markup. This can apply right down to a detailed feature markup.

By coincidence perhaps, identity of two human signals is determined by speakers and listeners not by minute and detailed examination of waveforms but by comparison of the symbolic representation of those signals. Thus the two waveforms in Figure 12.1 are each labelled [houwaju] (orthography: *How are you?*) and felt therefore to be the

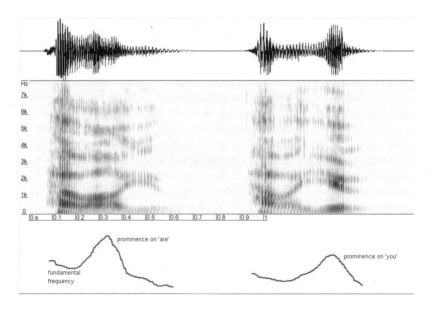

Figure 12.1   Two examples of *How are you?* In the left image the question is asked with prominence or focus on *are*, but in the right image the prominence is on *you*. Note how there are not just changes of amplitude associated with the shift of prominence but also differences in the whole utterance dynamics–resulting in relatively different timing and coarticulatory effects. This is a quite simple example of how expressive content has major effects on speech, making the choice of units in unit selection systems difficult, and posing serious problems for calculated transitions in many parametric text-to-speech systems.

same, despite the fact that objectively the signals are actually different. Speakers and listeners operate on *representations* of physical objects, not the physical objects themselves. It is for this reason that a pivotal focal point in our models of production and perception is the utterance plan: the *shared* object the speaker can report intention to say, and that the listener can report having heard.

In the more sophisticated systems, markup of the database indicates detail of the hierarchy of the data structure. Thus both segmental and prosodic markup is organised in terms of tiered structures. Usually segmental and prosodic markup are organised as separate data structures, but in our own model the two are unified. At its simplest, we consider that the prosodic data structure provides the environment within which the segmental data structure is organised; but in fact we go further in providing links between the various tiers for direct interaction (see Part IX).

Part IV

Areas for Improvement

13

Filling Gaps

13.1 General Prosody

Current synthesis may be satisfactory for many purposes, but there are some serious and important gaps in the various models in use. Many of these can be identified and efforts made to fill them. Areas which need attention vary from system to system, but there are a few obvious gaps or inadequacies which are shared by most systems. The most obvious and troublesome of these at the moment concerns prosody. The problem extends from the basic prosodics which is often associated with speech with minimal expressive content, through to treating the full range of subtleties of expression.

Prosody refers to intended phenomena in speech which span more than one segment. These phenomena include

- stressing patterns associated with individual words and combinations of words in phrases or sentences
- rhythmic structure of utterances
- intonation patterns.

The terms 'stress', 'rhythm' and 'intonation' come from phonology, and as such are associated with the cognitive processing human beings undertake to plan the prosody of their utterances. They are therefore abstract, and to make sense in a synthesis system have to be given physical correlates which are involved in the rendering process–the enactment of utterances plans. Here is the process in outline:

PHONOLOGICAL PROSODY		PHONETIC PROSODY
stress, rhythm, intonation	$\longrightarrow$ rendering process $\longrightarrow$	amplitude / timing / fundamental frequency (f0) contours adjusted appropriately

The physical correlates of stress, rhythm and intonation are variations in signal amplitude, timing fundamental frequency and occasionally spectral tilt. The correlation is nonlinear and variable; this makes the achievement of a satisfactory rendering algorithm for prosody

Developments in Speech Synthesis   Mark Tatham and Katherine Morton
© 2005 John Wiley & Sons, Ltd.   ISBN: 0-470-85538-X

problematical. We fall back here on our claim that ultimately we are interested in listener reaction to synthetic speech, meaning that arriving at a satisfactory rendering algorithm involves a great deal of perceptual testing and experimentation, rather than a straightforward analysis of the acoustic signal produced by human beings. Thus an understanding of how to render the underlying plans regarding stress, rhythm and intonation involves an understanding of how the acoustic signal is perceived. Nonlinearities in the perceptual process only compound the issue.

13.2 Prosody: Expression

All synthesis systems include prosodic processing, but this is almost always directed toward what is often called 'neutral' prosody. Prosody is used to communicate expression–a marker of how a speaker feels about himself or herself, who is being addressed and what the speaker is talking about. Neutral prosody is said to be prosody which excludes expressive content, and as such is often regarded as the carrier on which expression is to be modulated. Although this is a common enough analogy, there is some reason to believe that is unhelpful in that it implies that the carrier is able to be used without modulation–that it is possible for a speaker to produce an expression-free prosody.

It is probably the case that no human speech is ever without expression, though the intensity of the expression might be slight. It is perhaps for this reason that synthetic speech without expression seems very unnatural; listeners are simply not used to expression-free speech, and the acoustic signal immediately betrays the fact that it is synthesised.

Over the past few decades a great deal has been written by psychologists and phoneticians about expression in speech, or more specifically about how emotion is conveyed by the acoustic signal. Researchers have been examining the acoustic signal to find what it is that communicates to a listener that, for example, the speaker is happy or upset. Unfortunately although some very broad cues are clear (anger, for example, often involves increased rate, widened fundamental frequency distribution and increased amplitude) no satisfactory algorithms have yet been published for making a synthesised voice convincingly convey human expression. This is a pity because, as we explain elsewhere, this is perhaps the biggest remaining unknown area in understanding both the characteristics of human speech and synthesising a convincing simulation. It could also be argued that expression and emotion are two of the most significant properties of human speech, and that to omit them robs the signal of its humanness. This is precisely why synthetic speech rarely sounds completely natural.

But there is a serious paradox here. We can clearly and usually unambiguously identify emotion in a stretch of human utterance, and we can copy (synthesise) that utterance to produce a synthetic waveform which is convincing enough to listeners; but we still cannot characterise how such a signal differs from another which conveys a different emotion even if the actual words used are identical. Many researchers have referred to this paradox, and many have tried to solve it, but without consistent success. And yet solved it must be if we are to truly simulate human speech production. Our own opinion is that eventually the solution will lie in an improved theory of speech production which takes in expression not as a bolt-on afterthought in the model, but as a central or enveloping characteristic of utterances. In Chapter 25 we discuss a detailed reworking of speech production theory which takes this alternative approach.

13.3 The Segmental Level: Accents and Register

On the segmental level there is a shortage of modelling of dialect variation. True, all synthesisers speak with a particular accent, and there are many that do not reflect what might be termed the 'standard' accent of a particular language. But what we are referring to here is the global modelling of accents for any given language (notice this is different from the global modelling of dialects, which includes the semantics and syntax of the language–accent modelling is for the most part confined to the phonology and phonetics of the language).

Accent modelling is no longer regarded as a luxury for synthesis, and there are some good practical reasons for wanting a synthesis system to be able to speak with different accents. For example, in telephone information systems it might be useful to have the system respond to a caller in the caller's own accent–accurate regional accents can be a very sensitive issue. However, there is an important theoretical reason also for wanting to have multi-accent synthesis systems. Human beings can and do switch accent. This does not mean that they switch regional accent at will, but it is a statement much more about how they switch between accents on a different axis: the educational or status axis. Almost all speakers of a language adjust their accent slightly along this axis depending on who they are talking to. So speech synthesisers, to be convincing, will have to do the same–this is part of what human beings do when they speak, so it is part of what must be incorporated in any simulation. This is not something which speakers do just occasionally–it is all-pervasive, just as speaking with expression is all-pervasive.

This is an important point. When speaking to members of the family the register or style adopted is often different from the one adopted when speaking to the boss or to a stranger. Speakers are usually not particularly aware of doing this, or at least are not particularly aware of the details of the variation in style adopted. Whatever we call the phenomenon–register, style or code 'switching'–it is an important part of human speech, a kind of dimension of expressiveness. It involves modification to both the segmental and prosodic aspects of speech. Some coarticulatory phenomena, for example, might be cognitively restrained by what the psychologists have come to call *intervention*, speed and segment deletion might be curtailed with a view to speaking 'more properly', and so on.

The point is that if human beings do this kind of thing on a regular basis then it is part of human speech behaviour, and as such has to be included in a theory of speech. Theoretical linguistics models will typically neglect this behaviour altogether because it is not usually seen as anything to do with the underlying global properties of a language (universal or language-specific) in the abstract domain in which linguistics operates.

Although inevitably all synthesised speech is in this or that accent, there has been no real solution offered to the need to be able to produce particular accents on demand. A synthe-siser built in England will probably have one of the many British accents, and a synthesiser built in the United States will have one of the many North American accents. What would be ideal, though, would be the ability for synthesis systems to switch accents (perhaps even languages) with the minimum of difficulty and with little reprogramming. As a practical example, a phone company may well seek to answer local queries with an interactive system which synthesises the local accent. Another might be a reading machine for those with sight impairment which can switch accents depending on the material being read out loud or the accent of the person using the system.

The difficulty here is that linguistics has little to offer in terms of a theory of accent. True there are some huge collections of data–recordings of the accents of this or that language– and there are many detailed accounts of the range of pronunciations recognisable as this or that accent. (See Wells (1982) for a virtually exhaustive account of the many accents of English.) But there is no real *theory* of accent which might form the basis of synthesising this range of alternative ways of rending a language in synthesis.

A theory of accent would account for several aspects of speaker and listener behaviour when it comes to dealing with accent. Among these would be:

1 Speakers produce utterances which listeners recognise as consistent renderings in a *particular* identifiable accent, so accents are internally consistent and repeatable.
2 Speakers and listeners recognise an accent as being a systematic variant rendering within the framework of a *particular* language.
3 Speakers and listeners recognise the utterance 'behind' the accent. That is, they can identify (e.g. by writing down in ordinary orthography) the same utterance spoken in different accents, and can readily disassociate the utterance from the accent.
4 Provided speakers and listeners are 'tuned in' to the accent (i.e. they have internalised a characterisation of it) there is seldom ambiguity caused by the rendering process. The underlying utterance–the one which can be written down–does not alter because it is rendered in a different accent.

There are a number of interesting points which emerge here and which are potentially contentious so long as there is no accepted coherent theory.

The main one is that there is a stable internal representation for both speakers and listeners which is independent of accent. Those who would disagree might cite their claim that the phonemic structure of accents varies, even if they are accents within the same language. The claim rests on the notion that what underlies speech is the same as a surface analysis of speech once it has been produced–a claim long since relegated in contemporary linguistics as being unhelpful for appreciating the cognitive processes involved in language. Any apparent patterning in the surface acoustic signal is quite trivial against an underlying complexity based on making a distinction between deep and derived symbolic representations and then characterising their nonlinear relationship with the final signal. There is no reason why apparent patterns in the acoustic signal should directly reflect actual underlying patterns in the production of that signal. Theories which cannot access the cognitive processing behind the surface signal can say little about what is involved in the production and perception of speech.

In the absence of an agreed coherent theory of how utterances are recognised *despite* accent, we propose that underlying an utterance (which must, of course, be in *some* accent) there is an accent-free representation. That is, there is every reason to assume that at some level it is possible to speak of an utterance without considering which accent it is to be rendered in. This is exactly parallel to the notion that it is possible to speak of the lexicon of a language and the regularities of its syntax etc. without reference to any particular sentence of that language; or to speak of the meaning of an utterance without reference to the emotional content with which a speaker delivered it. Unless we can factor out such properties we cannot understand how they can apply differentially to different utterances in different situations.

So, for example, just as a speaker can say a particular utterance in such a way to communicate that he or she is happy or angry with the listener, so the person can say the utterance to communicate that he or she comes from Wales or London. The accent and the emotion do wrap the utterance, in the sense that they are all-pervasive throughout that utterance; but they remain, in some sense, distinct. The working hypothesis is, therefore, that there is an English to be modelled which is accent-free and expression-free. It would be absurd, though, to try to synthesise this English and hope for a natural rendering. The concept of the model rests on the hypothesis that all accents can be derived in a straightforward manner from this underlying representation. Counter-evidence would *not* come from an observation that different surface data analyses differently–for example, that a different phoneme set might be needed for different accents–*but from the failure of any algorithm to perform such a derivation correctly.*

13.4 Improvements to be Expected from Filling the Gaps

If the gaps which we are suggesting contribute significantly to the current lack of naturalness in current speech synthesis were to be filled, then we can expect naturalness to be improved along the following lines.

- The rendering of an utterance's prosodics would be more complete and more natural sounding to the listener. At the very least, utterances would sound correct, if a little stilted and initially lacking in other than minimal expressive content. We question elsewhere whether it makes sense to speak of an utterance with no expressive content, but for practical purposes it may be useful to imagine a neutral expression which is as complete as possible as a kind of baseline against which to evaluate expressive content.
- Expression and/or general expressive content would be easily incorporated and managed, including rendering of emotional and attitudinal aspects of speech. From a theoretical perspective expressive content pervades speech because of its all-encompassing or wrapper status. It would make sense to move quickly from a stylised neutral prosody to one which truly incorporates expression. For a discussion of what *natural* means, see Part IX, and for the notion that expression *wraps* utterances, see Chapter 35. The model underlying the synthesis system would embrace an architecture amenable to rendering utterances within a clearly defined domain of expression. The rendering process would modify itself according to the overarching demands of planned expressive content. A problem encountered in those systems which are beginning to include expression–namely the difficulty of introducing expressive content in a time-varying way–would be overcome. Expression is not constant throughout an utterance; indeed it may and perhaps usually does change from word to word. However, the domain of expression is very wide (often greater than utterance length) and as such constitutes a relatively slow-moving phenomenon compared with prosodic and segmental movement within the utterance itself. Expression can be thought of as functionally equivalent to a low-pass filtered phenomenon underlying the mid-pass filtered effects of prosodics and the high-pass filtered segmental effects.

In our model, expression and prosody are handled using a common data structure or data structure architecture based on XML (see Chapter 25 and Chapter 36). Low-pass effects dominate mid-pass effects, which in turn dominate high-pass filtered effects.

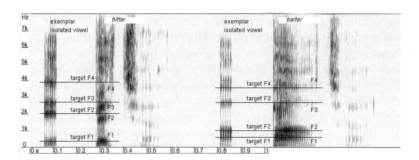

Figure 13.1 Two examples of exemplar target vowels–[ɪ] and [ɑ]–and how they appear in just one instance each of the words *bitter* and *barter* (British English with /r/ deleted). Note just to what extent the targets are 'distorted' by coarticulation with adjacent segments. The spectrograms show the first four formants of the targets projected on to the words, with the actually occurring, but dynamically shifting, formants labelled.

Some segmental effects will be better rendered, particularly those depending on accurate prosodics, like correct rhythm etc. Up till now the link between prosodic and segmental rendering has been largely neglected. But if the approach which sees segmental rendering as taking place within a prosodic domain is adopted, the link between segmental rendering variants and the current prosodic effect becomes clear. Without a correct prosody–reflecting a productive model of expression–the opportunity to improve these prosody-dependent variants of the segmental rendering is lost.

Segmental effects which may end up better rendered include phenomena like the accurate timing of coarticulation–effects which are often considered to be very difficult to achieve well. Coarticulation effects are usually agreed to be rate-dependent, with the argument that rate largely determines the extent of a coarticulatory effect. The idea has been that the greater the rate the more the effect will apply. Sometimes this just means that a target (the prescribed acoustic profile, in turn dependent on the target articulatory configuration profile) is held for rather less time than the duration of the entire segment; at other times it means that the target is missed altogether. Figure 13.1 shows the smoothing effect of inertial coarticulation.

However, it may well be that some effects are more rate-dependent than others. If this is the case then any cognitively sourced modification of the effect must use a rate-dependent index. We have known for a long time that cognitive intervention in human motor control and other processes–including apparently subconscious and involuntary ones–is normal (Ortony *et al.* 1988; Scherer 2001), but incorporation of the principle into synthesis systems is uncommon. Concatenative systems, particularly those able to process relatively long (i.e. greater than 'phone' length) segments or unit selection systems will be able to include many of these effects as a by-product of the use of large databases of stored human speech; but what is at issue here is the ability of the system to vary these effects as the prosody unfolds.

It is clear to us that the rate dependence of coarticulatory effects does not simply stem from some fixed inertial coefficient for each mechanical parameter which can be the object of coarticulation. There are variables at play here which must be appropriately taken into account. Even after stripping away any cognitively sourced constraint on coarticulation, we have to seriously consider the possibility that we are not dealing simply with an underlying fixed coarticulatory phenomenon but a variable phenomenon.

> The implementation of accents and dialects will undoubtedly improve the naturalness, and certainly the acceptability, of the synthetic output of systems which can incorporate a suitable model of accent in their higher level processes.

14

Using Different Units

14.1 Trade-Offs Between Units

In principle, different types of unit can be used for concatenation in any of the various low-level systems. The units are different with respect to their linguistic type, an abstract categorisation, and, in practical terms, their physical duration.

It is important to distinguish between linguistically motivated units and other units. One reason for this is that if a unit has linguistic motivation there are all sorts of properties associated with the unit which might be useful to us in the synthesis process. For example, if the unit is a syllable we know about its internal hierarchical structure and that it plays a pivotal role in prosody; the corresponding demi-syllable does not have these formal properties. Similarly, phone segments can be defined in terms of their physical features (these are not their abstract phonological features) that enable them to be classified in particular ways and consequently share some processes; the corresponding diphone would not behave in the same way.

However, there is sometimes a clear advantage to basing the system completely or partially on units which are *not* used within linguistics. This kind of situation might arise, for example, if some particularly difficult processing is needed for the linguistic units or if the system fulfils a niche need which does not call on the full complement of properties which might arise from having a linguistically derived system. For example, if our synthesis system is simply to pick up phrases from a list using a slot-filling methodology (as, for example, in announcements on a rail station or air terminal), then the largest units which will do the job are called for–not because they are linguistically relevant but because they avoid problems like coarticulation processing associated with the smaller linguistic units. Careful structuring of units can build in this type of processing in limited domain applications.

14.2 Linguistically Motivated Units

There are a number of different possible linguistic units, listed here arranged hierarchically according to abstract *length*. The term 'length' refers to two things in linguistics:

1 an abstract impression of what the duration of the acoustic signal might be once the unit is rendered, and
2 the hierarchical arrangement of contained units (a syllable contains one or more phonemes, a word contains one or more syllables, etc.).

Developments in Speech Synthesis   Mark Tatham and Katherine Morton
© 2005 John Wiley & Sons, Ltd.   ISBN: 0-470-85538-X

- *Whole utterance.* We use 'utterance' to mean a complete stretch of spoken speech, planned or synthesised as a single unit or within a single domain. The utterance may include one or more sentences, or be less than sentence length. Perhaps in our abstract system the best place for utterance is at sentence or greater than sentence length. The utterance is *defined* as a spoken unit, rather than as a grammatical or phonological unit; often, though, the utterance may *correspond* to a phonological domain or syntactic domain.

- *Paragraph.* In current research it is not quite clear how to define 'paragraph'. Is this a syntactic or a semantic unit? It could be semantic in the sense that it is a set of consecutive sentences collected for a semantic reason–a consistent 'theme'. Some researchers will define paragraph as a set of sentences–a syntactic definition.

- *Sentence.* A 'sentence' is a syntactic unit. In contemporary theoretical linguistics it is taken as the basic or focal unit for the theory. It is the longest domain over which the normal grammatical (or syntactic) rules operate. In our model it is the basic syntactic domain of the plan. Like all these linguistic units this one is abstract–which means we do not, at this stage, have to take into account phenomena such as incomplete sentences. When we come to physical language at a later stage in the model, though, we shall still need it as a reference point–the sentence which is *not* incomplete–against which to measure or identify any sentence which *is* incomplete, for example.

- *Phrase.* The 'phrase' is a syntactic unit, of which there may be one or several within the sentence domain. Some researchers identify phrases in terms of phonological prosody. Syntactically defined and phonologically defined phrases may or may not correspond (Zellner-Keller and Keller 2002).

- *Sub-phrase.* The 'sub-phrase' is a syntactically determined unit within the phrase. As with phrases, sub-phrases may be defined in terms of phonological prosody, and may also correspond to syntactic units.

- *Word.* 'Words' are syntactic units–they are the basic units which are manipulated within the sentence domain by means of the rules of syntax (the language's grammar). Words may consist of one or more morphemes; they constitute the grammar's terminal vocabulary.

- *Syllable.* 'Syllables' are primarily phonological units. They can be identified within words–words are either monosyllabic or polysyllabic. There are many monosyllabic words in English, with the average probably between one and two. Syllables are often thought of as the basic unit for phonological prosody, rather than the sub-syllabic segment (phoneme or allophone).

  - Notice that some researchers think of syllables as consisting of allophone units. But since some allophonic units are derived from phonemes within a syllabic environment, it follows that the syllable and its constituents must come prior to the allophone. We can recognise, then, an underlying syllable and a derived syllable.

- *Phoneme.* Still on an abstract level, 'phonemes' are sub-syllabic units. Syllables must consist of at least one phoneme (and this must be a vowel *nucleus*) to exist. Phonemes are the unpronounceable units which underlie the processes involved in building the phonological plan for an utterance. These processes may concern individual phonemes or, better, phonemes as units which exist within the hierarchical (rather than linear) structure of syllables. Another way of putting this is to say that syllables exhibit a hierarchical structure for their internal units–their constituent phonemes.

The above units relate therefore in a hierarchical data structure which we can characterise in XML in the following way:

```
<utterance>
        <sentence>
                <phrase>
                        <sub-phrase>
                                <word>
                                        <syllable>
                                                <phoneme/>
                                        </syllable>
                                </word>
                        </sub-phrase>
                </phrase>
        </sentence>
</utterance>
```

In practice not all utterances will contain all elements, and some elements may appear to overlap. Thus in the exchange:

speaker 1: *Who did it?*
speaker 2: *Me.*

the word *me* is, in a sense, *all* of–syllable, word, sub-phrase, phrase, sentence and complete utterance.

Before moving on to a-linguistic units which are often used in speech synthesis, we can use the opportunity to remind ourselves of two more units: the phonological *extrinsic allophone* and the phonetic *intrinsic allophone* (see also Chapter 11).

- *Extrinsic allophones* are the phoneme-sized units (i.e. they are dominated in the data structure by phonemes) which are used in the utterance plan. Since the utterance plan comes logically after the phonology, all phonological derivational processes have been traversed; these allophones include cognitively sourced variation (though not cognitively modified non-cognitive variation–see intrinsic allophones below). Since they are units of the plan they are abstract.
- *Intrinsic allophones* are phoneme-sized units (i.e. they originate from phonemes) which are used for the symbolic representation of post-coarticulatory variants in the utterance. Like extrinsic allophones they are abstract, though they reflect coarticulatory derivational processes. Intrinsic allophones also reflect cognitively sourced modification of coarticulatory effects.

14.3 A-Linguistic Units

One confusing aspect of terminology in this field is a not infrequent double or ambiguous use of terminology. One or two of the units we listed above as being well principled within a linguistic framework are *also* used with variant meaning within an a-linguistic or non-linguistic framework.

- *Utterance.* Although of considerable importance within our linguistically based model, the term 'utterance' is also used when no linguistic motivation is intended. In such a context it simply means 'what is spoken', and is rather informally defined.
- *Phrase.* The 'phrase', in this context, is an informally defined unit. It is less than an utterance in length, or less than a sentence if syntax is taken into consideration. A phrase is often taken to be a stretch of speech expected to be preceded or followed by at least one more other phrase(s) as part of a sentence to be spoken by a speech synthesis system.
- *Sub-phrase.* A 'sub-phrase' in the informal a-linguistic context is simply part of a phrase.
- *Demi-syllable.* The 'demi-syllable' is a loosely defined element used in some concatenated waveform systems. It is defined as a stretch of waveform running either from the centre of a syllable (see below) to the centre of the next syllable, or running between approximately steady-state periods near the centre of a syllable. There are no demisyllables in linguistics.
- *Syllable.* The 'syllable' itself in such systems may be equally informally defined. Syllables are needed linguistically as units of prosody, and in the a-linguistic usage are there to indicate areas of stress in the rhythmic patterning of an utterance.
- *Diphone.* 'Diphones' are defined either as a stretch of signal from the centre point of a phone (see below) to the centre point of the following phone, or a stretch from the least varying (most stable or steady-state) part of a phone to a similar point in the next phone. The idea of introducing diphones was to capture the transition between phones within the acoustic model, without having to model the transition itself explicitly. There are no diphones in linguistics.
- *Phone.* A 'phone' is a speech sound for many researchers. That is, it is an acoustic instantiation of a phoneme or allophone. Phones are usually considered as being able to be pronounced in isolation–that is, rather like 'This is what it would sound like if there were no coarticulation with the adjacent phones'. Useful for illustrative purposes, the phone is not, in our view, very useful in speech synthesis–not least because it is so informally and so variously defined.

Compare the phone, however, with the well motivated *abstract target* representation in phonetic target theory or the representations used in some early formant synthesis models (cf. Holmes *et al.* 1964, and others). One very real problem with the phone is the reluctance of researchers to specify exactly how abstract the unit is. For us it must be abstract if it is the 'sound of the target' or some such.

The hierarchical data structure relating the above a-linguistic units is as follows:

```
<utterance>
        <phrase>
                <sub-phrase>
                        <demi-syllable_syllable>
                                <diphone_phone/>
                        </demi-syllable_syllable>
                </sub-phrase>
        </phrase>
</utterance>
```

This is not an exhaustive list of a-linguistic units, nor is it the case that every tier in the hierarchy has to be represented. Where unit names are the same as a linguistic name it means that these are roughly equivalent terms; but in the a-linguistic system these are very often essentially derived or rendered physical units, whereas in the linguistic system they are underlying or abstract units.

14.4 Concatenation

Since one of the major problems in all low-level systems is how to treat the boundaries between abutted segments, it follows that minimising the number of occurrences of boundaries is likely to improve the output speech–certainly in this respect. There may be other ways in which a longer unit is likely to cause a drop in quality (particularly in the area of prosody), so it is essentially a question of trade-off. The smallest possible unit has no internal length (abstract) or durational (physical) boundaries. Other units which stand higher in the hierarchy already have any internal boundaries modelled correctly by definition–that is, when they are used, for example, in the marking of connected speech in databases. The term 'model' is used to mean that the waveform itself is used as an inexplicit model, and that its use obviates the need for an explicit model.

Reducing the number of boundaries involves, of course, using longer units. The point is that the longer the unit the greater the number and detail of boundaries contained within them. This, of course, is of no help to theorists because it is a clear attempt to avoid explicitly characterising boundary phenomena. There is probably no doubt that this approach improves the quality of the synthetic speech at the segmental level, but this simply underlines the problems with current theory. If we get better results with opaque models than we do with explicit characterisations, then the explicit characterisations are either wrong or being interpreted wrongly. It would be hard to disagree with the general notion that it is better to engineer a system with the support of good sound theory. But it would be equally hard to disagree that good systems are unlikely to emerge from bad theory.

14.5 Improved Naturalness Using Large Units

Joins between synthesis units are important. How critical a join is from the point of view of the perception of any errors depends on a number of factors, and there has been insufficient research to come to any firm conclusions. The research itself will be difficult to do conclusively, because perception is not a straightforward process. For example, if the perception of semantic content in the signal is easy, then it may well be that greater errors in the acoustic signal can be tolerated. If the perception of the semantic content turns out to be difficult for some reason (and this could range from conditions like environmental noise through to lack of familiarity with the topic), then errors in the acoustic signal may become more critical.

A first guess might suggest that the longer the unit length the less troublesome will be any errors–because larger 'amounts' of semantic content will be captured by each unit. If this is the case then errors in conjoining small units like phones will be the most critical for perception. This is precisely the reason why some researchers prefer diphones as the small units rather than phones–the very structure of the model is designed to minimise listener awareness of error, and errors are most likely to occur at coarticulated joins between segments.

Diphones, at least in theory, model the joins well, but at the price of having to store quite large numbers of permutations of units and having not infrequent awkward discontinuities between them which need smoothing in a phonetically unmotivated way.

Ultimately most conjoining problems will be associated with coarticulation. Coarticulation is the smoothing of abutted boundaries between segments. The smoothing function is directly dependent on the mass and mobility of the articulators involved, and is time-dependent. Although noticeable to the listener at the acoustic level, the smoothing actually takes place at the articulatory level, and involves either

- the mechanics of articulator movement, or
- the aerodynamics of the system.

Inertia tends to smooth any sharp transitions–that is, those with high-frequency content. Since coarticulation is time-dependent, the slower the speech the less the smoothing effect when expressed either as a percentage time for each of the abutted units or expressed in terms of how the ideal values associated with the parameters of the target specification for a segment were missed.

Coarticulation as a serious consideration in articulation theory presents itself only in those theories which rest on the notion that speech is concatenated isolable phones. Any theory which does not rest on the idea of joined-up segments has no problem with joins, of course. Concatenated waveform systems based on large databases are hybrid: they model all joins within the original recording perfectly–because the database is the recording. But they have problems as soon as they introduce a break or hypothesise a boundary. The reason for this is that a 'cut' in the acoustic flow is often made during a period of coarticulation, and any re-joining to different excised elements must, by definition, result in an error at the join. Syllable-based systems–for example SPRUCE (Tatham and Lewis 1999)–make their cuts only at syllable or word boundaries, where it is hypothesised that there is likely to be

- less critical coarticulation
- less perceptual focus (so less *detected* error).

SPRUCE uses syllables partly for this reason, and also because in its underlying theory the syllable is the focal unit for prosody which is implemented in accordance with the 'wrapper' theory (see Chapter 25).

The idea of using larger units is simple. If the critical coarticulation effects are between individual phones (as articulatory or acoustic abstract targets based on underlying phonemes which are deep symbolic representations of the eventual instantiations), then the larger the unit the greater the number of joins will be modelled without further processing. For example, following on from the discussion above, we can use different low-level speech synthesis systems, resting on different symbolic analyses based on sizes of units ranging through

- phonemes
- extrinsic allophones (planning units)
- syllables
- words

- sub-phrases
- phrases.

If, for a given utterance, the number of coarticulation points which needs to be modelled is around p (we shall not give an exact number to avoid discussion of some of the more complex arguments of phoneme and allophone theory), this number is reduced to around $p/2$ if the unit is syllabic (if on average a syllable contains 2 allophones), or around $p/4$ if the unit is the word (if on average a word contains 1.5 syllables), and so on. The probability of noticeable error correlates with this reduction, all things being equal. However, much of the gain from choosing larger units is purely practical. Avoiding modelling each join avoids developing further the theory of coarticulation for the purposes of synthesis–which is a loss to this particular application of phonetics.

But avoidance of coarticulation in this way stores up an unexpected problem. The degree of coarticulation varies according to many different factors, only *one* of which is linear phonetic context. To have in the database inventory of units the syllable *speech*, for example, does not guarantee that the four phones conjoined here will always be accurately joined when the syllable occurs in a new context in a newly synthesised utterance. Since coarticulation is rate-dependent, the syllable will be intrinsically correct only if it is used for a new sentence in precisely the same way it occurred for the database recording–which is highly unlikely. This does not mean that adjustments cannot be made to compensate for departures from the

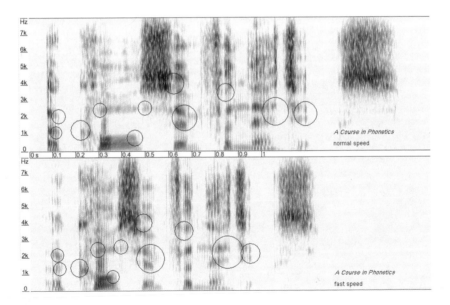

Figure 14.1 Two examples of *A Course in Phonetics*, spoken by the same speaker (with apologies to Peter Ladefoged). The top example is spoken at a normal rate, while the lower example is spoken a little faster. Areas of interest showing differences between the two utterances are circled. Note that the whole 'internal' dynamics of the utterances are different, not just some overall linear variation of general rate. Listeners are very sensitive to these internal dynamics, and it follows that appropriate rendering is essential for the perception of complete naturalness.

original recording environment; it simply means that doing so increases the chances of error *and* requires us to have a good conjoining model *anyway*. Even more problems are likely to occur when it is realised that contextual coarticulatory influence can 'spread' to up to half a dozen segments on either side of any one segment. Figure 14.1 shows spectrograms illustrating how coarticulation varies under different rate conditions.

15

Waveform Concatenation Systems: Naturalness and Large Databases

Particular problems arise with waveform concatenation systems (including unit selection systems) when it comes to prosody and expressive effects. Waveform concatenation is especially effective at the segmental level, but is strictly limited at the supra-segmental level because the number of possible combinations of prosodic variants and segmental variants is very large indeed.

At the segmental level, a particular unit has to be considered in terms of its segmental context. This context might be taken into account on a linear basis as in the simpler models, or on a hierarchical basis in more sophisticated models. Take, for example, the contextual effects in the syllable. The simpler model involves just enough context to characterise linear relationships:

```
<syllable>
        <consonant_string/>
        <vowel/>
        <consonant_string/>
</syllable>
```

Here, coarticulatory processes in the phonetic rendering depend on which consonant precedes and which follows the vowel nucleus of the syllable.

The hierarchical model establishes a relational structure:

```
<syllable>
      <onset>consonant_string</onset>
      <rhyme>
             <nucleus>vowel</nucleus>
             <coda>consonant_string></coda>
      </rhyme>
</syllable>
```

Here, coarticulatory processes depend on the relationships established within the underlying hierarchy. Interactivity between the initial consonant string and the vowel is different from that between the vowel and the final consonant string.

Developments in Speech Synthesis   Mark Tatham and Katherine Morton
© 2005 John Wiley & Sons, Ltd.   ISBN: 0-470-85538-X

This means that if our units are phonemes (phonological units) or physical phones (phonetic units), then our attention will be focussed on assimilatory and coarticulatory effects respectively. The database markup will have to include these contexts such that, for example, in a pair of word sequences found in the database such as

the stop
the top

the markup for /t/ will have to indicate that, on rendering (and considering its place just within these two words), it coarticulates with [s] and [o] in *stop* (with no delay in the onset of voicing for the [o], but that it coarticulates with [...] and [o] in *top* (with delay in the onset of voicing for the [o]. These [t] sounds in the database have to be marked as segmentally different. In creating a new synthetic utterance the appropriate [t] will be located, say for the words *storm* and *toll* respectively. The segmental markup and selection can be made pretty exhaustive provided we limit assimilation and coarticulation to a single 'depth' and consider only linearly adjacent segments. The idea dates back to a contextual theory in the 1970s which only a minority of phoneticians espoused; it was proposed by Wickelgren (1969), a computationally oriented psychologist, and involved hypothesising that ready-made contextualised segments were stored in the speaker. An utterance was produced by employing the completely contextualised segment from the store, rather than allowing coarticulation 'on the fly' as the utterance was rendered.

Thus the phrase *speech synthesis* might in the commoner coarticulatory model be marked as a string of permutable extrinsic allophones:

s p i ʧ s ɪ n θ ɪ s ɪ s

but in Wickelgren's model as

$_{\#}S_p$ $_sP_i$ $_p$ˈtʃ$_i$ ʧ$_s$ ʧ$S_ɪ$ $_s$ɪ$_n$ $_ɪ$n$_θ$ $_n$θ$_ɪ$ $_θ$ɪ$_s$ $_ɪ$S$_ɪ$ $_s$ɪ$_s$ $_ɪ$S$_{\#}$

Note that the storage requirement in Wickelgren's model is very large compared with the extrinsic allophonic model. It can be seen that for the phrase *speech synthesis* the coarticulatory model involves seven distinct symbols, whereas in Wickelgren's model the same phrase involves twelve—every major symbol needs as many entries as it has contexts on either side (including boundary contexts, indicated here with the symbol #), since Wickelgren is, in effect, attempting to mark all immediate coarticulatory contexts within each composite symbol.

Very few, if any, contemporary phoneticians subscribe to Wickelgren's model for human speech production, but in sharp contrast the implicit model in synthesis systems which select units from a large database is *almost precisely* the one Wickelgren proposed. The underlying motivation, of course, is that contemporary theory in phonology and phonetics prefers to allow for maximising computational possibilities to provide for the nuances of variability, whereas the unit selection philosophy is to capture as much as is thought necessary of this variability within the stored database. The models could not be more diametrically opposed.

For those interested in phonological and phonetic theory, we can mention two factors that militated against Wickelgren. First, there is the size of the database and the problem of 'on the fly' retrieval—a contemporary preference was for runtime computation rather than retrieved stored lists or large-scale data. Second, consider evidence from speech errors, especially 'metathesis' where elements are transposed, as in

a cup of tea becomes *a tup of kea.*

Here, the final acoustic signal gives us the expected coarticulation dependent on the final segmental context and *not* the original one–whereas Wickelgren's model would have predicted that segments moved *with their original context intact* and did not become re-embedded in their new contexts.

15.1 The Beginnings of Useful Automated Markup Systems

It might not be necessary for markup to indicate coarticulation and other similar phenomena, such as the comparable effects which occur with prosody. It is probably possible to deduce these, and this is the route taken by most researchers so far. The problem is that the resulting description has perhaps been too simple so far. Only the grosser aspects of coarticulation will be able to be deduced and subtleties will not be obvious. The observation made earlier about the different [t] sounds could probably be deduced by rule–except that it may vary according to expressive or even a fairly bland or so-called neutral prosody. It may be possible to consider the trade-off between marking and rule application. We have to ask ourselves whether we have enough knowledge to re-mark by rule, as opposed to marking up databases by hand.

One of the reasons why this is important is that automatic markup is thought by many researchers to be an important task when databases are to be very large–which they must be to include as many examples of as many variants as possible. Various approaches are possible:

- markup by hand
- simple automatic markup augmented by hand
- simple automatic markup augmented by hand, but with the system learning from the hand intervention
- fully automatic markup.

Markup by hand, as we have remarked, is probably the most accurate, but the least possible with large databases. Augmented automatic is probably the most favoured approach at the moment, but an automated system which can learn by interaction with human intervention is clearly a strong contender for advanced systems. Fully automatic markup is not a realistic option at the moment since not enough is known about the relationship between symbolic marking and the acoustic signal to develop the required algorithms–hence the need for hand intervention. It is to be hoped that a learning device will eventually make up for this lack of knowledge.

15.2 How Much Detail in the Markup?

Having asked the question whether coarticulatory phenomena at the segmental level should be included–or indeed the equivalent variation in prosody–the next stage is to consider just what depth of markup would actually be useful. When it comes to indicating on the database an appropriately aligned symbolic representation, we can see that two types of segment conjoining effect have so far been considered possible.

- *Assimilation*. This is the introduction of segment-element variability dependent on *phonological* context. This context can be modelled linearly, or, more successfully, hierarchically; the markup is in terms of the extrinsic allophones derived in the utterance plan.
- *Coarticulation*. This is the markup of segment-element variability dependent on *phonetic* context. This type of variability can also be modelled either linearly or hierarchically; the markup identifies intrinsic allophones derived during the utterance plan rendering process.

Synthesis models such as the Holmes *et al.* (1964) text-to-speech system include a rule-based system for calculating both types of variability; the rules are all context-sensitive rewrite productions. Unit selection systems usually mark the database with as much variability (of both types) as can be accommodated, given the available level of automated markup.

But there is scope for going to a greater depth, and a revised approach is going to become more important as we proceed to requiring sophisticated levels of expressive content in our synthetic speech. So how might we achieve greater sophistication in the markup? Paradoxically one way may be to actually mark less on the waveform in terms of actual surface or derived acoustic or linguistic detail. The reason for this is that any one stretch of speech excised from the database for inclusion in a novel utterance may need *further* modification. This is likely to occur as we include more and more expressive content. Vast databases with every nuance of expression occurring on every possible combination of prosodic and segmental elements are not the answer.

Large databases with marking indicating candidate areas for expressive nuance together with a generalised set of procedures for dealing with expression are much more likely to emerge.

In our suggested overall model of speech production we make provision for cognitively sourced constraints on coarticulation–an otherwise purely physical phenomenon. These constraints are overseen by the 'cognitive phonetic agent' (Tatham 1995; Tatham and Morton 2003) acting in a supervisory role. It so happens that we intend the same type of approach to be responsible for how cognitively sourced expressive content is introduced. There seems to us to be no reason why short-term coarticulatory effects and longer term prosodic effects should not be modelled in a similar fashion. The principle is that there are phenomena which can be modelled as neutral to *external* constraint, but which can also succumb to external constraint in a principled and carefully supervised way. So far successful speech synthesis relies on being carried by implementing these neutral effects, such as plain phonetic coarticulation and phonological prosodics. In principle these neutral effects are abstractions, and we would argue that they can never actually occur without constraint of some kind. We have argued at length that it is the introduction of these constraints precisely to depart from plain synthesised speech which leads to restoring the naturalness now felt to be lacking in synthetic output.

The neutral approach, based on the prosodic wrapper model

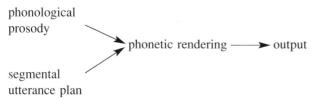

The neutral approach, extended to include supervised expressive control

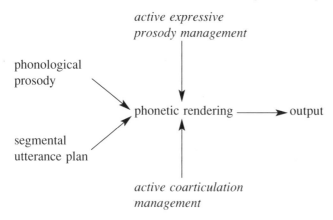

The above illustrates the basic supervised expressive approach which is designed to enable CPA active management of rendering processes which would otherwise lead to an unnatural neutral output. The neutral output is not dissimilar to abstract exemplar sentences obtained by running a static syntax–useful from a theoretical perspective, but not suitable for effective or convincing rendering.

The illustration is simplistic since it begs many questions, not least the exact point at which cognitively supervised intervention takes place; but it serves here to make the point that both prosodic and segmental aspects of speech output are probably the result of managed variability within the rendering process. This managed variability has 'meaning' for the listener, if only to indicate that the source of the acoustic signal is a human being. But because the variability is predictable only from the *current* pragmatic context, it cannot be stored in the database underpinning the synthesis system. It must be generated 'on the fly' in a strictly controlled manner. For precisely this reason we make the following suggestion.

The database should be marked to indicate points in the prosodic and segmental structure which are open to further controlled expressive variability. These points must be suitably marked to enable a set of procedures *external* to the database to introduce the required variability as the final utterance is being rendered for output. We believe that this strategy will not only lift the perceived degree of naturalness in the acoustic signal, but do so in a way principled by an appropriate model of human production of expressive content.

So, it is not a question of marking ever more detailed information in the database, but a question of indicating where more detail can be introduced from a source external to the actual synthesis run; and then, of course, providing the means to do so by setting up procedures which can be called to modify the data structure. There is provision within an XML-based markup to do just this.

Thus segmental and prosodic markup interact to make this happen, as when degree of coarticulation may vary depending on expression. Sometimes, for example, clarity and precision are called for even at rates of utterance where they would normally assume less importance. In angry speech people often speak faster, yet with more precision, apparently contradicting one of the basic ideas in coarticulation–that the degree of coarticulation correlates with rate of delivery. It is interesting, though, that even when under some circumstances prosody and segmental rendering do usually interact, they may not always do so. Similarly some aspects of precision of articulation may be tied to rate, but not always. These are complex issues and as yet researchers in speech production are unable to give us all the details; the experiments have yet to be done. It is important, though, that we establish an appropriate framework for dealing with phenomena of this kind if we are to make progress in developing improved speech synthesis.

15.3 Prosodic Markup and Segmental Consequences

Notwithstanding the issues raised above, there remain one or two basic problems for prosody, particularly in concatenated waveform systems. Within contemporary synthesis which does not attempt expressive use of prosody there are two opposed bases for handling prosody.

1 The existing prosody of the database can be normalised out with a view to completely recreating it for new utterances.
2 The database prosody can be left intact and, if the database has sufficient examples of each segment with varying prosody, tapped for supplying the prosody for the output utterance.

The second approach is theoretically unsound because it assumes that prosody is an attribute of the segment.

15.3.1 Method 1: Prosody Normalisation

Several systems have adopted the *first method* involving normalising out whatever prosody was present in the recorded database. One of these is the SPRUCE system, which, since it focuses on high-level synthesis in detail, leaves little to chance when driving a concatenated waveform low-level system. In SPRUCE the prosody is developed to the point where expressive detail is included. Note that in the alternative approach, the combinatorial problems associated with finding every possible nuance in the database would, especially for expressive detail, make a system unworkable. In such a system, the ultimate database would need to

- *reflect* every segmental possibility before coarticulation
- have examples of every coarticulatory possibility in basic, enhanced and constrained versions

- have examples of all of these possibilities in any possible position in all prosodic envelopes
- have all expressive permutations of prosodic parameters for each possible envelope.

We do not need to put actual numbers on these propositions to guess that 'huge' would be a word grossly underestimating the size of the database required.

The SPRUCE literature does discuss, however, that the method loses the detail of micro-prosody and that this is probably perceptually important. This can be overcome by marking (before normalisation) the points where micro-prosody will occur–and omitting these areas from the normalisation. The problem is, of course, that the parameters of prosody and micro-prosody are the same (hence their shared name), and in doing a blanket normalisation on the one we are in danger of affecting the other. In practical terms this means taking perhaps two or three cycles of fundamental frequency after certain plosives, for example, and not applying prosodic normalisation to them. The question then would be whether this leaves too disjunct a rendering once the general prosody is applied. It would be necessary, of course, to make sure that the synthetic prosody did not overwrite the values for these two or three cycles. It is not algorithmically difficult to modulate a new prosodic contour with the preserved fragment of micro-prosody.

Marking and retaining micro-prosody before general prosodic normalisation is completely sound from a theoretical perspective, since, contrary to what might be expected from its name, micro-prosody is not really a prosodic phenomenon. Prosody is about how the small number of phonetic prosodic parameters (fundamental frequency, amplitude and rhythmic timing) change over time domains greater than segment length. Micro-prosody is about how these same parameters–but now not involved in prosody–change as a coarticulatory phenomenon between adjacent segments. The idea is not dissimilar to the way in which (cognitively determined, not coarticulatory) changes of fundamental frequency within the syllable domain constitute phonemic tone in a language like Mandarin, or allophonic contrastive tone in a language like French. Just because fundamental frequency is involved it does not mean that we are within the prosodic domain. Figure 15.1 illustrates the micro-prosody phenomenon.

So, micro-prosody aside, in the first option for developing prosody from a recorded database a phonetic prosody (consisting of putative f0, amplitude and rhythm contours) is computed and then the retrieved string of units is fitted to this contour by adjusting each unit to fit within the contour. The major problem associated with this approach is the accuracy of the entire prosodic model–usually computed as a two-stage process, abstract (phonological) representation followed by physical (acoustic) representation.

15.3.2 Method 2: Prosody Extraction

In the *second method for developing prosody* the problem is generally seen as one of marking the different prosodic features for each identical segmental feature–thus considerably increasing the number of required different segments, or, with a fixed database, considerably reducing the number of repetitions of the 'same' segment. To a certain extent micro-prosodic or intrinsic effects will be already modelled, particularly if the smallest chosen unit is the syllable; but the ability to find a precise match to the planned segment is compromised by the size of the database.

It is hard to see how this technique could work really satisfactorily. The actual fundamental frequency values of the reassembled segments cannot match up properly, unless the

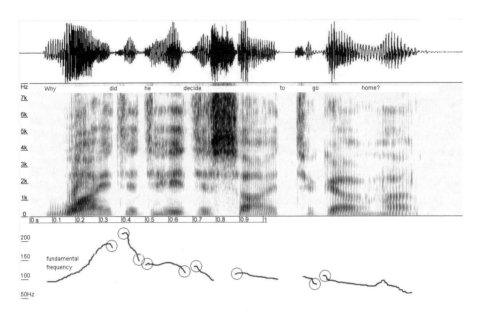

Figure 15.1   The utterance is *Why did he decide to go home?*, spoken with the normal falling intonation associated with *wh*-questions. The waveform, spectrogram and, in particular, the fundamental frequency curve show the effects of microprosody–circled. Although this is micro-intonation, one type of microprosody, it must be remembered that it is neither intonation nor prosody in either phonological or phonetic terms. These are local aerodynamic coarticulatory effects which distort phonetic prosody, but are not part of it. Microprosody is appropriately modelled as a completely different phenomenon from prosody proper. However, it is a phenomenon (like all coarticulation) to which listeners are particularly (if subconsciously) sensitive. It is essential that it be modelled appropriately in synthesis to achieve perceived naturalness.

database is vast and the segments sub-phone in size. A smoothing process should iron out many of the discontinuities, but some micro-prosody might be lost, especially with very small segments. The basic technique of using the intrinsic prosody of the database for reassembly of units is theoretically unsound in the perspective of an overarching prosody. It assumes that the optimal way of modelling the relationship between prosody and segments is to make the prosody fit the segments, even though it is computed beforehand at a higher level.

In any case, reproducing the acoustic prosody from the database is as counter-theoretical as reproducing segments from the database, and subject to exactly the same kinds of theoretical criticism. The answer is going to be that so long as things work on a practical basis then it doesn't matter. In turn, the answer to this is that, laying aside practical gains, we loose the opportunity to evaluate any theoretical approach we may have adopted, or to test our theoretical approach.

It is clear that some researchers may not be aware of the problem of dealing with detail in prosody. Take, for example the almost extreme poverty of ToBI-style markups (see Chapter 30) when it comes to displaying detail. While probably satisfactory for a neutral prosody within a phonological framework, this kind of markup is not at all compatible with the detailed markup of a real database. And to force the markup is to court problems when it comes to rendering detail. The abstract markup cannot reflect the actual fundamental frequency values, only relative values. So a fall, for example, from 150Hz to 125Hz might be labelled identically to a fall from 200Hz to 165Hz (to say nothing of the *shape* of the fall), which may be crucially different from the perceptual viewpoint. As far as the selection algorithm is concerned, either of these examples would constitute a hit and it would be hard to choose between them without differential marking.

15.4 Summary of Database Markup and Content

We can see that, as with many aspects of speech synthesis and automatic speech recognition, there are trade-offs to be had between different approaches. Much research has been devoted to determining the optimum approaches–but clearly a dynamic approach which can *vary* its dependence on just how much is modelled *within* the database compared with the intended utterance plan will, at least in theory, eventually score higher in terms of intelligibility and naturalness of output.

- Is it really feasible to have some kind of hybrid approach which seeks to extract abstractions from the database and then *re*-synthesise the acoustic signal (introducing errors during the process) and which also takes *actual* stretches from the database despite the fact that in theory the retrieved instantiation can *never* be exactly what is needed?
- How would a hybrid system actually work?
- How would the constantly shifting dependence on intrinsic database content *vs* computed utterance plan actually be managed?

Whatever the approach and the final solution in any one system, it will be necessary to compute a high-level abstract representation of the prosodics underlying the intended utterance. Because of this, the huge database unit selection systems still fall short of producing a completely natural output. The problem lies not just in the size of the database but in the sophistication of the high-level model. It is fairly clear that what we have available at the present time is not fully adequate on either front–the underlying representation and the database content (together with its markup).

The high-level model of prosody needs to be able at least to

- identify areas of the database which match up with the available symbolic representations
- avoid ambiguity in the symbolic representation which is eventually applied to the database
- mark in such a way that the actual acoustic signal in the database constitutes a plausible instantiation of the markup–it can never be the *actual* instantiation
- ensure that the markup and the *apparent* plausible instantiation must also be a possible markup and *plausible* instantiation of the new utterance or part utterance being synthesised.

These matters are far from trivial, but do underline the need to be as explicit as possible regarding the differentiation between abstract symbolic representations and the set of instantiations they characterise. By definition, no two instantiations will ever be identical, and no instantiation contained in a database can ever be the required instantiation for a novel utterance. Ultimately it is a question of compromise; but the compromise has to be informed and based on pre-determined perceptual thresholds for error in detail.

16

Unit Selection Systems

16.1 The Supporting Theory for Synthesis

Just as much when dealing with unit selection synthesis systems (Black and Campbell 1995) as with other types of low-level synthesis, we are concerned with the actual quality of the underlying theory at this level and particularly at the dominant higher level of synthesis.

> Is the theory on which the system depends up to the job of providing the basis for either the database prosodic markup, or for predicting the prosodic contours of the new utterances to be synthesised?

At present the answer is *no*, and this is largely for historical reasons. Contemporary linguistic theories are intended first and foremost to be descriptive of what speakers and listeners *feel* about the prosody of utterances and the segmental utterance plan fitted to that prosody. The models are psychologically oriented, and were not therefore originally designed to predict the prosodic detail of the acoustic signal. The reason for this is quite simple–variability in human speech is extremely complex, and is characterised by processes introducing it at a number of different levels, some of them extrinsic and voluntary and others intrinsic and involuntary.

At a cognitive perceptual level the symbolic representation assigned to an acoustic signal by a listener is notable for its *lack* of variability, hardly squaring well with the variability of the acoustic signal itself. The exact relationship between the apparently relatively invariant linguistic/psychological descriptions and the very variable acoustic signal is not known in sufficient detail to enable us yet to build accurate models which can predict the variability in an acoustic signal from an invariant plan. It is also questionable whether current prosodic models are yet sufficiently good to provide the invariant plan itself, let alone move on from it to the signal itself.

Researchers often speak of criteria properties of the symbolic representations in their models using terms like *level*, *fineness of detail* or *granularity*. Traditionally phoneticians have used terms like *broad* or *narrow* of transcription (the symbolic characterisation of classical phonetics) to indicate the degree of detail in phonetic intended to be represented. One of the problems with these characterisations of the markup or transcription is that they are

Developments in Speech Synthesis   Mark Tatham and Katherine Morton
© 2005 John Wiley & Sons, Ltd.   ISBN: 0-470-85538-X

quite inexplicit. We can try to be, as a first approximation, a little more explicit in the way
we go about using such terms.

16.2 Terms

Borrowing from phonological theory, we can identify the ultimate underlying or maximally
abstract level–and this we have termed in this book *the phonemic level* which identifies phoneme
units. We define phoneme as the name of a class of variants, which, when used symbolically as an element, dominates those variants in a functional way, optimised for representing
the underlying sound patterning of morphemes in such a way as to distinguish them uniquely.
All elements or symbols below that level incorporate some representation of derived variability. It is clear that for maximum clarity we need to identify appropriate levels as we
proceed from the phonemic level toward the actual acoustic signal, and be clear at each
level just what is the variability incorporated in the description.

Several researchers have identified various types of variability introduced between the
ultimately neutral phonemic or highest level in the phonology and the lowest level–where
we find utterance plans. They have also identified similarly various types of variability introduced between the utterance plan and the final acoustic signal, this time derived during
phonetic rendering of plans. Following investigations by various researchers in the 1970s
and 1980s (MacNeilage and De Clerk 1969; Fowler 1980; Tatham 1986a,1995; Tatham and
Morton 1980; Lindblom 1990) the models are relatively stable and the distinctions between
phonological and phonetic processes adequately agreed.

For the most part this idea of tiered processing has so far been applied to segmental
representations, but the idea is relatively recent in supra-segmental or prosodic representations, though the beginnings of a formal model can be seen, for example, in Firth (1948).
Researchers like Pierrehumbert (1981) and the proponents of the ToBI markup system have
provided the framework for reasonably comprehensive characterisations of abstract phonological prosody. Pierrehumbert and others have provided ways of rendering abstract phonological markup in terms of acoustic contours, but the results are far from satisfactory and
often look like transposed phonology, owing to their lack of variability, rather than genuine
phonetic rendering. A phonological markup of prosody needs to be sufficient to lead to a
formal declaration of

- underlying prosodic contours analogous to underlying phonological segmental (phonemic)
  representations–this set of contour data structures needs to be able to uniquely characterise distinguishing prosodic features and no more (i.e. no variants);
- processes or sets of processes which derive all possible prosodic utterance plans within
  which segmental phonological plans fit;
- derived prosodic structures exhaustive of the possibilities for wrapping all declarable
  segmental plans–any one prosody declared at this level is exemplar only, just as any one
  segmental utterance plan is exemplar only.

These three requirements are captured on the phonological *static plane* (see Chapter 22
and Part VII). Phonological rendering processes exist on the phonological *dynamic plane*
for deriving, for any one sentence, one prosodically wrapped utterance plan expressed as a

string of extrinsic allophones, ready for phonetic rendering. Thus an instantiated prosodic-ally wrapped phonological utterance plan is declared at the exit of phonology.

16.3 The Database Paradigm and the Limits of Synthesis

Typically in unit selection systems, the database is marked up in ways *dependent* on such abstract descriptions derived from the linguistic models–though what is being marked is an acoustic signal is far from abstract or invariant. The act of markup involves an inexplicit data reduction process which purports to arrive at a symbolic representation higher up the structure which derived the actual acoustic signal. A markup of this kind declares what might have underlain the acoustic result. In typical systems it is inexplicit because what is 'left out' is often not clear or consistently justified. More often than not the markup is worked up in a reverse fashion, beginning with the acoustic signal. At its very best such markup constitutes a hypothesis about some symbolic representation which *could* underlie the actual acoustic signal found in the database.

The fact that the backwards derivation is inexplicit is demonstrated by the failure of automatic systems to provide markup as consistently useful as hand marking. At the present level of conceptualisation any markup is bound to be inexplicit and contain an *ad hoc* element because it fails to recognise all possible sources of variability in the signal. In particular, the reverse engineering (for that is what it boils down to) of a neutral markup from pragmatically driven explicit content is problematical. Many proponents of large database synthesis systems try to make sure that the recordings are of speech which contains as little expressive content as possible; and this is fine, provided it is recognised that there is ultimately no speech that does not have *some* expressive content (Tatham and Morton 2004).

Markup in unit selection systems must entail a vocabulary of symbols (or names of elements) which matches what high-level processing in the system will derive. Obviously a mismatch between the terms of the utterance plan and the elements used in the markup would be unfortunate. Partly for the sake of simplicity–essential for optimal robustness–the models minimise variability. We have explained this practice in database markup, and the match-ing lack of variability in the accompanying high-level processing is largely toward the same end: robustness. Some early researchers misunderstood linguistic theory, and equated invari-ant exemplar utterances derived on the static plane of phonology with actual utterance plans, and this led to a certain amount of confusion.

But one way or another–either because of the idiosyncrasies of the supporting linguistics or because of the demand for simplicity–variability is minimised in both the declaration of the novel utterances the device is to speak and in the database markup which will supply the necessary acoustic units.

16.4 Variability in the Database

The difficulty is that the database itself *does* contain the variability, and this is precisely why the database approach has been so widely adopted. It automatically contains the variability which may guarantee a natural synthetic output. But systems rarely if ever incorporate the means of getting at this variability–the means of access to it have been formally denied in the data reduction process inherent in the symbolic markup. Thus, for example, a useful number of variant renderings of, say, a syllable might appear in the database marked identically, and thus preventing optimal choice. There may be ways around this, but there is an inherent theoretical contradiction here. So far, all that has mattered is a non-optimal use of the available *neutral* naturalness, so systems do not sound as natural as they might. But if the demand for non-neutral expressive content increases any further, we shall be obliged to introduce markup systems at odds with what we have at present–ones that deliberately focus on expressive variability in the prosody and the derived segmental variability which is entailed by the dominant prosodic wrappers in the original recording. To repeat: most if not all of this useful (and in the future, critical) variability is currently lost to data reduction and minimalist derivational processes underlying the utterance plan. A reversal of the current strategy could involve a major upheaval.

At the same time, during the synthesis process itself the utterance plan is usually created from the same or similar models, resulting in a description with little or no variability present. The utterance plan, remember, is the final symbolic representation of what is to be spoken; it constitutes the output of a dynamic phonology and the input to dynamic phonetic rendering in human speech production. The utterance plan will match well with database segments marked up minimally, but on both sides of the equation (before and after the pivotal utterance plan) the variability so characteristic of speech production is either ignored or negated in an unhelpful way, leading to listener appraisal which is bound to note poor naturalness. The obvious conclusion is that the naturalness judgement depends at least in part on the detection of variability–not because variability is used in decoding the message or even crucial in decoding the speaker's expression, but because listeners expect a variability they do not find.

As a parallel, consider the so-called script fonts available on personal computers. If you sign your name using one of the fonts no one will believe that this is really your signature; it may look like handwriting but everyone knows it is not. The parallel is not exact, but unless the script front is capable of displaying variability analogous to coarticulation and individual variations, it will not be confused for real or natural handwriting. A whole paragraph written in this stylised or idealised font will be tedious to read. Similarly stretches of speech, or stretches of dialogue, rendered in an idealised form lacking in variability will be tedious to listen to.

16.5 Types of Database

Early concatenative systems used databases of pre-selected units, such as phones, diphones, syllables or demi-syllables. The creation of these databases involved selecting a single unit for inclusion in the system–representative of all possible occurrences. The difficulty was in obtaining a single all-purpose representative version of the unit in an *offline* selection process; that is, the units are pre-selected. To cater for all possible segmental or prosodic contexts, the unit needs to be as neutral as possible. The paradox of course is that the more general purpose the unit the more abstract it is and the less natural. A phone-based system provides the worst situation, with just a hundred or so units.

Unit selection systems are at the other end of the scale. Here the idea is to include as many non-neutral units as possible, and choose the appropriate one *online*–that is, the units are selected at runtime; hence the need for efficient search techniques. Segments excised in the selection process vary in size: some systems opt for diphones, with some going as far as triphone length segments. The excised units do not always have to be the same size, of course, so *dynamic* selection can involve varying length units–that is, sub-phone up to *n*-phone units. Unit size measurement is usually in terms of phones, where a phone is a section of audio signal labelled symbolically using an intrinsic or a contextualised extrinsic allophone.

The labelling unit is important. On the one hand the selection is driven by the utterance plan, probably expressed as a string of extrinsic allophones; but on the other hand the database itself consists of audio corresponding to what we would usually think of at the symbolic level as intrinsic allophones–coarticulated extrinsic allophones. So either the symbols are appropriately contextualised extrinsic allophones (i.e. allophones indexed with their context), or intrinsic allophone symbols already incorporating the context (see Chapter 19).

Whichever model is chosen (unit lists or a database of running speech) there are two terms, *unit* and *context*, involved. The choice is between adding context (in the case of small isolated unit database systems) or storing contextualised units (in the case of large database systems). So, typically:

phone/diphone ⟶ select ⟶ compute and add ⟶ output in context
database   unit   contextual effects

or

contextualised database ⟶ select unit ⟶ output in context

Whether segmental and prosodic contexts and their effects on units are computed at runtime or whether the units are stored already contextualised, it will not be possible to have available all possible units in all possible contexts. We need either

- an exhaustive model of variability and its sources in order to be able to add context (for example, diphone + context rules), or
- a very large database already containing units in all possible contexts (e.g. unit selection).

Clearly neither is completely feasible. The more extensible of the two approaches, and also the one which more closely matches what we think happens in human beings, is the approach which computes and adds contextual effects at runtime.

The use of diphones or demi-syllables was originally intended to get around the problem of context, just as the large database systems aimed at avoiding the problem. But context is much more complex than the simple surface linear effects of the early coarticulation model (Öhman 1966; MacNeilage 1970) on which both solutions are ultimately based. In particular, both systems fail to consider contexts larger than the immediate segmental context, and fail to take prosodic context into account. So far, however, unit selection has proved superior in terms of perceived segmental quality and naturalness, underlining the weakness of conjoining models, and weaknesses in the theories which characterise speech as a string of conjoined isolable segments.

16.6 Database Size and Searchability at Low-Level

16.6.1 Database Size

The size of the low-level database–the actual recording of human speech to be used for deriving units in the synthesis process–is a major factor in determining naturalness in the final synthesised signal.

At the segmental level the coarticulatory permutations of units runs to many thousands; and when each of these can be an instantiation within a different prosodic environment, the number is multiplied up many times. It is not clear what the minimum number of instances of any one segment or prosodic effect would be ideal to ensure naturalness, although some researchers have tried to estimate this. There are suggestions that the minimum size of a database should consist of at least 1 hour of recording, with some researchers suggesting 7 or 8 hours.

We could estimate that if the number of diphones necessary to express just once each pair of possible co-occurring segments in English is 1500 (the usually quoted number), then to have just one example in each of, say, three stress levels would be 4500. If there are six different prosodic contours averaging strings of (say) 30 segments in any combination, and ten different types of expressive use of each of these contours, and (say) at least three different rates of delivery for all possible combinations of the previous elements, then there is no need to do the calculation to make the point–the combinatorial problem is literally astronomical. It is as ridiculous to suggest that natural human speech can be represented by a database as it is to suggest that a language can be characterised by listing its sentences. In both cases we are dealing with an unarguable and absolute impossibility, because both databases would have to be infinite in size.

> But, we could argue, there is no call for representing every possible utterance in speech production and no call for representing all sentences in syntax. Theoretically this argument too is unsound, if only because perfect representation is simplicity itself for every human being. And human beings *clearly* do not do things by enumerating all possibilities in advance! The language and speech of human beings is *not* database-driven. While this remains true, perfect speech synthesis is not possible on this basis, and some other means must be developed to achieve high-quality synthesis. The theoretical position is so robust that there is no tenable argument for basing synthesis which is to truly simulate human speech production on the database paradigm. We are aware that this is a strong statement.

In terms of how *human beings* produce speech, the generalised database or unit selection model is unhelpful and wasteful of scientific intellect. However, as a practical interim solution to producing some kind of synthetic speech which is certainly virtually 100% intelligible and often fairly natural sounding, it is enjoying quite considerable success. But it is clear there are diminishing returns in terms of improved naturalness against database size, computational load and intellectual investment.

Plainly the model is incorrect, though equally plainly the linguistic/psychological model currently used in the non-database speech synthesis is *also* incorrect or we would never have had recourse to the database approach. There is an interesting parallel in automatic speech recognition: clearly a statistically based approach, like that offered by the popular hidden Markov model (HMM) approach is *not* how human beings assign symbolic representations to the acoustic signals they hear. But it works well enough–or at least works better than the state-of-the-art modelling of the human psychological strategies involved.

So we fall back to a more tenable position, either because perfect synthesis is impossible, or because we decide we do not want it after all. But, having cleared the ground, we are in a much better theoretical position than before. The trick is to begin listing what we do *not* want our synthesis to do. Synthesis is a maturing technology *and* a maturing science. In the pre-science period we did what we could and settled for barely adequate techniques and results. But considerations like *Is my synthesis intelligible?* are for the past, not the future! We are done with the belt and braces era.

The most natural speech possible has appropriate and perceivable

- expressive content
- semantic content
- pragmatic content.

It has been encoded properly using agreed coding conventions characterised by

- semantics
- syntax
- phonology
- phonetics;

. . . as well as special strategies characterised by socio- and psycho-linguistics which deal with expressive and pragmatic content in detail. Knowledge of the properties of language and the chosen language in particular result in a *rendering* which enables decoding sensitive to the original

- expressive content
- semantic content
- pragmatic content.

The rendering strategies are such that they work with perceptual strategies to handle error correction and permit us to incorporate unusual effects–such as, to give a trivial but striking example, tell jokes based on incongruity of content and/or rendering.

16.6.2 Database Searchability

Vast databases require very efficient means for searching them, especially if speech is to be generated in real or near real time. The search engine itself needs to be particularly efficient, but also the markup of the database and the corresponding marking of the planned utterance which drives the search needs also to be efficient and point quickly to the required units. Characterisation of the utterance plan and the way the database is marked need, of course, to be strictly complimentary.

Part V

Markup

17

VoiceXML

17.1 Introduction

The development of speech technologies such as automatic speech recognition and text-to-speech synthesis continues to offer an as yet unrealised potential for use in a wide range of voice-oriented applications. Many of these are telephone-based and will at some point in the future offer a genuinely useful alternative to human-based systems such as call centres. But despite decades of promise these technologies have not yet reached the stage of sophistication where they can be used on a reliable basis, or command full user acceptance.

Despite the fact that this might be true–and although the promised horizon seems to approach painfully slowly–it is important for researchers and developers to begin to think in terms of how the eventual integration of these technologies might work. How, for example, will the structure of an information system integrating speech recognition and synthesis be implemented in a useful, transparent and even extensible manner? How will these technologies integrate with allied technologies such as machine translation or document retrieval? The message here is that the time is right for effort to be spent on optimising a holistic approach to such systems. It is too early for rigid standardisation, but it is not too early to be investigating what might need to be standardised at some stage in the future when the technologies are ready.

A number of integrating or overarching technologies have been proposed or are undergoing development. We choose as an example of such a technology VoiceXML (Sharma and Kunins 2002; W3 Consortium 2004), partly because it is in use already in a few areas and partly because it is relatively robust. The assumptions on which VoiceXML is based and the way it conceptualises voice-based applications are coherent enough for it to provide us with a useful example of what will eventually become commonplace.

VoiceXML is not about synthesis or recognition of speech *per se*. It is the basis for a formal integration of these technologies with each other and with other allied technologies. Its strength lies in its approach to dialogue management which is uncomplicated. If VoiceXML does nothing more it will enable application designers to discover gaps in their prior conceptualisation of how dialogue works. Linguists and psychologists will take care of the basic theory of language and dialogue; artificial intelligence specialists will take care of the theory of how dialogue flow is managed in human beings. But dialogue application authors will discover what is needed from these technologies to make them work together in a real-world environment. The work need not wait since all the technologies are sufficiently developed for toy and, in some cases, real applications to be attempted and mistakes to be made.

Developments in Speech Synthesis   Mark Tatham and Katherine Morton
© 2005 John Wiley & Sons, Ltd.   ISBN: 0-470-85538-X

VoiceXML is a technology oriented toward enabling applications, and taking the form of a markup system deriving from XML (W3 Consortium 2000; Kim 2003). VoiceXML uses XML markup with a particular objective in mind–providing the document-based architecture for applications which focus on human/machine dialogue and which include audio. The dialogue can include the usage of all or any of the following:

- *input* using automatic recognition of speech and DTMF (dual-tone multi-frequency–touch tone), general voice input telephony, recording of spoken input
- *output* using synthetic speech or stored digitised audio.

With little effort, mixed media input and output might be included in an extended specification. Thus VoiceXML can be regarded as essentially an *integrating technology*. It is important not to confuse strategies and mechanisms for integrating technologies with the technologies themselves. These are explicitly external to VoiceXML.

17.2 VoiceXML and XML

The way VoiceXML conforms to the XML general specification makes sure that XML tools can work with VoiceXML, and in particular that XML parsers can be used to validate VoiceXML documents for syntactic well-formedness. Platform independence is central to the concept, and the focus on web-based applications able to deliver content and voice interactivity is a keystone of the implementation. A VoiceXML document, like an XML document, takes the form of a plain text file appropriately marked up with the available elements and their attributes to form a fragment of a voice-based application. The use of the terms 'text' and 'document' constitute an anachronism stemming from the original but historic purpose of 'marking up'–namely to specify page (paper or screen) layout. In many applications of XML this original conception of markup is still appropriate (Sunderland *et al.* 2004), but the concept is marginalised in favour of much wider meanings for such notions as markup and document. We shall continue to use these terms because they are current, but we mean them in their broader contemporary sense.

When specifically web-based, VoiceXML documents are associated with the now traditional semantics (logical flow) of web applications. Thus HTTP protocol is used for the retrieval of documents, and client-side logical operations readily include native support for JavaScript and other local processing. In principle the entire range of technologies available on the web from advanced streaming media through to simple cookies is at the disposal of VoiceXML for incorporation into voice-oriented applications. In addition, support is available to enhance interactivity with users in the form of a rudimentary set of telephony call control options. From within a VoiceXML document the semantics of various call transfer innovations (including call disconnecting) contributes to an unprecedented package for versatile application in dialogue systems.

17.3 VoiceXML: Functionality

VoiceXML provides the framework for managing a group of wide-ranging technologies into human/machine dialogue applications. What it does *not* do is specify, constrain or improve on the ability of these technologies to perform their designed functions. Within the scope of this book, for example, VoiceXML does not directly improve the quality of synthetic speech or the success rate of speech recognition–it provides the means to include seamlessly in a single application what synthesis or recognition is already available to the developer. Standalone applications become integrated *via* this technology, and even more importantly, become integrated with the *human* user. This is the main significance of the approach.

Below is a summary of VoiceXML functionality:

1 INPUT

- recognise speech (manage automatic speech recognition device)
- recognise DTMF (manage touch tone recognition)
- record audio input

2 OUTPUT

- deliver pre-recorded audio (manage digital signal processing device)
- synthesise speech (manage text-to-speech device)

3 GENERAL

- process HTTP semantics for document transactions
- process general forms and include client-side scripting
- telephony control (including call transfer)

17.4 Principal VoiceXML Elements

The top-level element wrapping all VoiceXML documents is <vxml>. The structure of the document within the wrapper is relatively simple and is based on a small number of elements, some of which declare general properties while others specify well-defined actions. The following is a basic selection from among the available VoiceXML elements. Bear in mind that most elements can be, and sometimes *must* be, accompanied by qualifying attributes. This is not a tutorial text on the use of VoiceXML, so these attributes have been omitted here for the sake of conciseness.

- **<form>** is the basic element used in interactive applications. Within the form, data can be collected, modified, acted upon and delivered.
- **<field>** declares an input field within a form. The input field enables user input, often for subsequent action.
- **<filled>** enables an action to be executed when fields are filled usually by the user.
- **<noinput>** traps a no_input event.
- **<nomatch>** traps a no_match event.
- **<prompt>** queues speech synthesis and audio output to the user.
- **<block>** wraps a section of executable code which usually does not involve interaction with the user.
- **<if>/<else>** is an associated pair of elements enabling basic conditional logic.
- **<goto>** redirects flow execution to another identified dialogue or sub-dialogue in the same or a different document.

- **<grammar>** specifies a grammar for speech recognition input, or a strategy for decoding DTMF (dual-tone multi-frequency) input. The grammar is often to be found in the specified external file, which itself can be in XML format.
- **<help>** traps a help event.

Not all elements in VoiceXML are relevant to the use of speech synthesis. Indeed, VoiceXML documents might well contain reference to simple interactive dialogues in the form of pre-recorded audio. The basic range of elements relevant to synthesis is very simple and rather restricted in the amount of control it allows the document author. Remember, the document is in effect a generalised script for guiding telephone-based interaction; these synthesis elements are either information or instructions to the attached text-to-speech system.

- **<prompt>** sets the audio files or synthetic utterances to be used as output in the interactive dialogue.
- **<audio>** plays the named audio file(s) or synthesised utterance(s).
- **<voice>** specifies the voice the text to speech system is to use, or the voice of the audio file to be played. This is the mechanism for allowing the caller to interact with more than one recorded or synthesised speaker.
- **<say-as>** expands text which is not fully orthographic, such as dates and abbreviations. Use of this element can either augment or override expansion strategies held within most text-to-speech systems (see below for the use of this element in speech synthesis markup language–SSML).
- **<div>** encloses text that is a paragraph or sentence (the general case).
- **<paragraph>** encloses a paragraph.
- **<sentence>** encloses a sentence.
- **<phoneme>** encloses text that is a phonetic transcription which, if necessary, overrides that generated by the attached text-to-speech system. The aim is to add variation to the synthesis which the text-to-speech system has no provision for, or when it is known to generate an error.
- **<prosody>** determines the prosody for the enclosed text.
- **<emphasis>** stipulates that the wrapped text is to be synthesised with emphasis.
- **<record>** records user spoken input.

Some element names are represented in full *and* abbreviated forms–for example, <emphasis> and <emp>, and <prosody> and <pros>. These together with the pair <say-as> and <sayas> create a redundancy some authors may find more confusing than it is helpful. Another anomaly which, on the other hand, could have benefited from being elaborated is the use of <audio> to specify a block of either pre-recorded audio or synthetic speech. The restriction of <audio> to a recording and the introduction of an element like <synthesis> might have been more helpful if not more logical. Such a usage would keep the document author aware of the fundamental differences between the two possible outputs, often reflecting a major qualitative difference as far as the interactive user is concerned. This is precisely the kind of thing which authors need to be constantly aware of in optimising the structure and content of their documents. Note that in VoiceXML, in particular, and in SSML (discussed below in Chapter 18), use of words like *prosody* or *phonetic* may be rather different from our usage in this book. One of the problems continually encountered in the field of speech technology is variant usage of technical terms.

17.5 Tapping the Autonomy of the Attached Synthesis System

One or two of the elements provided by VoiceXML bridge the gap between the autonomous behaviour of the attached speech system and what is going on in the dialogue flow itself. This is for the moment a weak area of VoiceXML. It is important that some autonomy be maintained within the synthesiser, but the synthesiser needs input *on a continuous basis* which is otherwise quite unpredictable and which follows from the nature of the unfolding dialogue. Thus, as no more than a general example for the moment, the pragmatic input required by a modern synthesiser is a variable which often stems from an evaluation of the progress of the dialogue (see Chapter 37 for explanation of the use of the term 'pragmatic information' in speech synthesis). VoiceXML is not yet equipped to provide this information. We shall find the same problem with SSML which is weak in the area of prosodic control.

In terms of the general model we propose: it is VoiceXML (or one of its successors) and SSML (or one of its successors) that will need to have the following properties:

- sensitivity to dialogue flow–especially the *developing* relationship between the human and machine participants, and
- the ability to turn this information into appropriate pragmatic and expression markers for *wrapping* the synthetic output.

A simple example will be enough here. Consider a dialogue between two human beings. One is asking for information, and a second feels she is providing it. But the first repeatedly fails to get the point, getting more and more frustrated as the second becomes increasing irritated. The speech of both clearly reflects their changing feelings. This is not an unusual situation, and certainly fledgeling automated information systems are highly likely to experience the problem. The system has to be able to detect the way the relationship is developing and control how its own voice reflects *its* attitude. How irritation is expressed is known to the synthesis system, but *when* to use it and *how* to develop it is information which must come from outside and which moreover is variable and time-sensitive: it *develops* and is not static.

18

Speech Synthesis Markup Language (SSML)

18.1 Introduction

Speech synthesis markup language is an XML-based markup system which has as its object-ive making available a relatively standard way of controlling and manipulating aspects of speech being synthesised by an attached synthesis system. The full specification is available from the World Wide Web Consortium (W3 Consortium 2003). In principle, any speech synthesis system can take control from an SSML document, though it must be able to under-stand the commands. Thus compliant systems will tend to be levelled in the sense that their range of functions will tend to standardise around those dealt with by SSML. This gives SSML developers an important responsibility to avoid lowest common denominator situ-ations, and to *anticipate* future developments by incorporating transparent extensibility. It is anticipated that SSML can provide a finer level of control over many functions, thus contributing to an improved and more consistent output quality.

The usual distribution of tasks is adopted by providing elements, together with their asso-ciated attributes, to handle the structure of the document to be synthesised, and how text is to be processed and pronounced. Prosody, potentially enabling style and expression variants, forms an important subset of the available control elements. A few miscellaneous elements are included: these have a tidying up function for the most part.

18.2 Original W3C Design Criteria for SSML

The W3C design criteria for SSML are quite clear.

Consistency

The intention is that SSML control of synthesis should not be specific for any platform or synthesiser. In terms of platform this is readily achievable in the sense that SSML is an XML derivative, and XML itself is not specific to a platform. With regard to the synthesiser, this would only be the case if all synthesisers were equally compliant with the SSML specification, and this is unlikely to be the case unless the synthesis system is explicitly designed to be *fully* SSML compliant. For the moment SSML itself might be consistent and adopted

Developments in Speech Synthesis   Mark Tatham and Katherine Morton
© 2005 John Wiley & Sons, Ltd.   ISBN: 0-470-85538-X

consistently into SSML documents, but the results of its control on the final synthesis can be no more than *approaching* consistency.

Interoperability

A number of web-based technologies could potentially interact with SSML. In this book we are particularly concerned with VoiceXML, and here mutual support is assured. Provided these allied technologies are developed bearing in mind the SSML specification, interoperability is achievable. W3C mention also audio cascading style sheets and Synchronized Multimedia Integration Language (SMIL™), which are allied technologies developed within the same framework.

Generality

Complete generality will be hard to achieve; the potential range of applications is far too wide. For example we shall see that a particular weakness of SSML is in the area of prosody control and how it relates to the synthesis of expressive content. Many simple applications will be tolerant of gross categories for expression (e.g. 'pleasant', 'authoritative'), but there will also be many which require considerably more subtlety (interactive medical diagnosis is an example) as yet not catered for. The weakness here lies less with the SSML specification than with the underlying speech production models on which synthesis rests—expression is usually barely touched upon and remains one of the last frontier areas of improvement for speech synthesis.

The ability of SSML to call on different voices from appended synthesisers is limited only by what is available in the synthesiser technology. Whether synthesisers can achieve different voices by transformation of existing voices or whether in some systems large alternative databases of recorded material are required is not *per se* a problem for SSML.

Internationalisation

Coverage of different languages (either within a single document or between documents) is not hard. It depends on whether SSML 'knows' of the existence of any particular language and whether the synthesiser being addressed has the required database of recorded units (in a concatenative system) or the segment specifications and transition rules (in a parametric system) for the required language.

Switching between languages is a trivial problem compared with switching between accents within a single language. There is no doubt that accent switching will be a requirement in sophisticated dialogue systems. For example, regional accent switching is appropriate for handling queries originating across an entire language area. This may seem for the moment to be just a luxury requirement, but TV advertising has shown the power of adopting not just regional accents but accents which reflect the educational or income brackets of the target audience. The problem is not dissimilar to providing 'young' responses (vocabulary, syntax, voice and accent) to use in systems where the expected inquiries are from 20-somethings or teens—a 50+ voice in an 'old fashioned' accent is just not appropriate and will kill some applications (see Chapter 13)

Generation and Readability

The idea here is that some documents will be hand authored–the markup will be provided by direct human intervention. In a large-scale or real-time environment, however, automatic generation of the markup is more appropriate. It is intended that SSML should be suitable for both automatic and hand authoring. For obvious reasons, like checking the output, any documents which have been marked up should be meaningful to a human reader. This is the same as saying that SSML should be a markup system using only plain text.

Implementability

Clearly to meet most of the above criteria the SSML specification should be able to be implemented with existing technology. The designers would ideally like to keep what they call 'optional features' to a minimum, thus ensuring an evenness of results.

18.3 Extensibility

A criterion which could have been included, and which we ourselves would deem essential, is a requirement that SSML be transparently *extensible*. Developments in speech synthesis strategies and the models on which they are based are likely to come fairly rapidly in the next few years as we learn which areas need focussed development. The example we return to repeatedly in this book is the need for synthesisers to tap expression as the route toward convincing naturalness of the synthesised voice. SSML probably is *not* extensive enough to handle the complexities of expressive content and will need considerable work in this area.

This is not just a question of adding more elements for marking finer details of prosody, for example, but possibly a question of a rethink of the hierarchical organisation of the markup system (see Part III for a discussion of how to introduce expression into synthesis). The theoretical details of the subtleties of expression (including style, mood, attitude and emotion) are not yet fully developed, but they will come and systems must be ready to accommodate them. That accommodation is likely to be in terms of the markup system used for the controlling document, since most synthesisers will be computationally able to produce the desired effects but unable to do so without a suitable control input.

18.4 Processing the SSML Document

There are various ways in which an SSML document can be created (e.g. automatically, by human authoring, or a combination of these), but it is the SSML definition which accounts for the form of the document. SSML is essentially a support device for text-to-speech systems, so the main burden of the rendering process all the way from text to goal acoustic signal rests with the intrinsic properties of any particular text-to-speech system. Rather than take over responsibility for general rendering, SSML is simply a means of making explicit specific pronunciation properties of objects within the document. SSML is conceived as having an adjunctive role, but in addition as a unifying system in as much as it is not confined to any one text-to-speech system. This notion of unifying systems within a specific relationship only works, of course, with text-to-speech systems that are explicitly compliant to the SSML definition. By 'unifying' we mean that an SSML document applied to several compliant text-to-speech systems will produce outputs more 'correct' and more similar to each other than would otherwise have been the case. SSML therefore brings similar constraints to those speech processors to which it is applied; their basic behaviour though is system-specific.

The W3 specification provides for six stages to processing an SSML document's markup content–that is, rendering the document for automatically generated or synthesised voice output. Each of these stages can be independently controlled from within the document by means of the way the SSML is applied to its content. This means that within the provisions of SSML there exists the means of dealing separately with each of these sub-processes. After the two initial stages, the four subsequent processing stages are roughly what most text-to-speech systems would usually have as their basic architecture anyway, and cascade in the usual order. Using the W3 terminology, the stages are

1 XML parse
2 structure analysis
3 text normalisation
4 text-to-phoneme conversion
5 prosody analysis
6 waveform production.

18.4.1 XML Parse

This initial and preparatory stage uses an XLM parser to extract the control document's tree and content, an action which potentially determines how subsequent processes proceed in detail. The parser is of a generalised XML type since SSML itself conforms to the overall XML specification.

18.4.2 Structure Analysis

Determining the document's internal structure is important because sections of the document which are to exhibit like behaviour are identified. The reason for this is that the synthesis processor which is responsible for actually creating the goal soundwave will need to be able to recognise like domains (stretches of text and subsequent utterance which constitute the range for certain sub-processes) to which to apply like strategies. For example, key domains in the satisfactory assignment of prosody are the paragraph and sentence–certain prosodic rules operate similarly on each paragraph in the document, provided it is of a specific type. In the same way, sentences of particular types attract specific prosodic rules which apply only to the sentence domain.

SSML elements dedicated to structure analysis are <paragraph> and <sentence>, used to delimit appropriate stretches of the document's content. If the synthesis processor does not receive these overriding markup elements it must proceed within its own structure analysis for those parts of the document which are unmarked. At all stages the synthesiser may have to rely on inbuilt procedures, and because of differences between devices there may well be outcome variations.

The document content (i.e. the text to be read out by the synthesis system) has a structure, which is independent of what is to become of that document; the text, for example, can remain unspoken and continue to exist as a parallel written document. In a synthesis system the structure can be made explicit by means of SSML markup, or be left to the synthesis processor to deduce the structure for itself. Within SSML the two elements, <paragraph> and <sentence>, are used to define basic document structure, but if SSML document structure is not provided the synthesis processor itself will use its own analysis procedure relying on punctuation etc. to arrive at the document's structure. It is also possible for a combination of explicit and derived document structures to be used. In this case SSML essentially defines a basic framework, and the detail of the structural description is dependent on the sophistication of the speech processor's intrinsic analysis procedures.

18.4.3 Text Normalisation

Not all representation in the orthographic system of languages is suitable for direct conversion to speech. For example there are numerical representations (like: 1, 345 etc. or III, MMIV) which will need an intermediate orthographic representation before rendering–in these examples, *one*, *three hundred and forty-five* (British English) or *three hundred forty-five* (American English), *three*, *two thousand and four*, and so on. Similarly abbreviations will need expanding: *Dr* can represent *doctor* or *drive*, *DC* can represent *direct current* or *District of Columbia*. Others will need just spelling out: *USA* is *U-S-A*, *CBS* is *C-B-S*. Almost all speech processors have one or two inbuilt sets of procedures for dealing with these situations, but none is perfect because of ambiguities inherent in the orthographic conventions.

SSML supports a means of enabling document authors to assist disambiguation by introducing the <say-as> element which enables a portion of text to be explicitly expanded into a representation without ambiguity. We can speak here of replacing a *partially specified* orthographic representation (which includes abbreviations and other specific symbols) with a *fully specified* orthographic representation in terms of appropriate strings of plain text symbols.

> The <say-as> element in SSML is dedicated to text normalisation and is aimed at making explicit how abbreviations and special symbols are to be pronounced. Most text-to-speech systems are quite good at text normalisation, but the <say-as> element is particularly useful either if it is predicted that the usual rules will fail or if the special text is a new abbreviation or symbol, or (in conjunction with the next stage) is an adopted foreign word.

18.4.4 Text-To-Phoneme Conversion

The orthographic representation of languages found in normal spelling is very rarely phonetic. There are several types of orthographic representation, but the two most relevant to our purposes are those which constitute a morpheme-based system (found, for example, in English and French among others) and a pronunciation-based system (as in Spanish or Italian). Rarely does a language use the system it is based on in an exclusive way–English *does* represent its own pronunciation to a certain extent, and Spanish *does* represent its own morpheme system to a certain extent. It may be more accurate to speak of a spectrum of preferred bases for orthographic representation.

One problem is that even pronunciation-based spelling systems fail when it comes to regional or social accent variations. Even in those languages whose orthography is usually considered to be a close approximation to representing their pronunciations, it is almost always the case that dialectal and accent variations eventually cause problems in directly interpreting the orthography in terms of pronunciation. Consider, for example, the differing typical Iberian and Mexican ways of pronouncing a word like *corazón* (respectively [kɔra'θɔn] and [kɔra'sɔn]). Inevitably there have to be procedures in a synthesis processor for converting orthographic representation into a fully specified underlying phonological representation. These will vary in detail and complexity depending on just how far removed the orthography is from not just the language's pronunciation, but the accent and dialectal variants which may be called for by the document's author.

The job of text-to-phoneme processing, then, is to replace the orthographic representation of the original document with a phonological representation which will eventually lead more transparently to the required pronunciation. We use the term *phonological* in its sense in modern linguistics–the representation is abstract and forms the basis for subsequent rendering as an actual acoustic signal. In Chapter 25 we explain why a hierarchically organised two-stage approach is found to be essential in linguistics, and why it is sensible to carry this model over into speech synthesis. Unfortunately most text-to-speech systems do not explicitly recognise this two-tier model, and for the moment this fact constrains SSML to an approach which is in our view less than optimum.

No synthesis processor can yet handle all the detail of orthography to phonological conversion. And perhaps this is hardly surprising when we consider that human native speakers often trip up over the conversion process when reading text out loud–particularly with unfamiliar words which cannot be deduced by analogy with other representations.

SSML provides for explicitly enabling words and other strings of orthographic symbols to be rendered directly as strings of what the SSML specification calls *phonemes*–strings of symbolic objects able to be subsequently processed in a variety of ways toward the final pronunciation goal. The W3 specification is inconsistent here–some of the provision is for the underlying units we ourselves prefer, but some is for a more surface representation, leading to confusion when subsequent processing occurs.

The author of an SSML document can provide a closer approximation to pronunciation than otherwise indicated in the usual orthography. This is done using the <phoneme> element or perhaps the <say-as> element, or the two in conjunction. In addition, sets of pronunciation definitions in the form of external lists can be referenced using the <lexicon> element–this might be found useful when the synthesis processor cannot itself derive different accents from underlying representations. Thus there might be separate lexicons for generic American or British pronunciations. The problem is confounded by the fact that there are numerous American pronunciations just as there are numerous British pronunciations, but this is a general problem not by any means confined to using SSML.

Once again, a hybrid approach is useful, tapping the synthesis processor's own system for converting orthographic representations to underlying phonological representations for subsequent rendering as specific surface pronunciations, assisted in difficult areas by explicit SSML markup. This will undoubtedly produce, if handled carefully, an improvement on the usual inbuilt systems. Explicit SSML markup would need to be sensitive to the dictionaries of words which do not conform to pronunciation rules which are normally found in text-to-speech systems.

18.4.5 Prosody Analysis

Prosody refers to *intonation*, *rhythm* and *stress* in phonetics and phonology. Intonation is perceived fundamental frequency change, rhythm is perceived structure in the timing of the speech signal, stress is perceived from local or short-term changes in fundamental frequency, amplitude and duration. All speech has prosody, and prosody is also the phenomenon which carries expressive content in speech. The prosodic system in speech is extremely complex, though some inroads have quite recently been made into understanding how it works (Tatham and Morton 2004).

SSML supports some very basic marking of prosodic features using the <emphasis>, <break> and <prosody> elements. However, SSML provides no detailed markup system for prosody, and consequently relies heavily on the synthesis processor's inbuilt strategies for assigning prosody. The procedure usually involves analysing the entire document structure, the syntax of paragraphs and sentences which form the document, and any other information carried by the text which may assist in deriving an appropriate prosody. Part II has more details of prosody in general and how it is managed in text-to-speech systems. In a sense there is a trade-off between the SSML markup and what is built into the synthesis processor; how this is handled is up to the system designer, and does vary from system to system. This is not always the disadvantage it may seem because it does enable the designer, with careful handling, to compensate for deficiencies on either side.

SSML elements are thought of as 'high-level' markers in that they handle patterning above the level of simple production of the acoustic signal. In linguistics terms we could say that SSML elements are mostly about what happens in the planning or phonological stages of subsequent instantiation of utterances. Elements like <phoneme> are responsible either for providing or refining actual content, and other elements relate to logical descriptions of style. However, the <break> and <prosody> elements are intended to kick in during later stages in generating speech. This means that they must complement the non-markup behaviour of the low-level system, and accommodate any higher level markup which has been applied. The way in which the interaction takes place must of necessity be specific to particular speech processors, since the latter vary in how they operate and in the level of performance expected of them.

Notice that SSML assumes prosody to be applied *after* many other processes have been determined by the system or marked for rendering by the system. Our own approach, on the other hand, places prosody much earlier in the hierarchy–using prosodics to wrap lower level utterances or parts of utterances (see Chapter 34).

18.4.6 Waveform Production

When it comes to the actual audio, <voice> enables selection between different voices enabled in the specific synthesis processor (so this is processor-dependent), and <audio> allows for the insertion of pre-recorded audio, provided this is allowed by the specific processor. Apart from the use of these two elements, the final stages of waveform production by the synthesis processor are left up to the processor itself.

18.5 Main SSML Elements and Their Attributes

SSML elements can be usefully grouped under the headings:

1 *Document structure, text processing and pronunciation*. These are the elements which participate in the first four stages enumerated in section 18.3. They are used for defining the document in terms of its overall structure and making sure that the synthesis processor can read the incoming text unambiguously, particularly when it comes to special symbols and abbreviations. Any anomalies of pronunciation and other special renderings are also taken care of by elements in this category.
2 *Prosody and style*. The elements dedicated to prosody and style are intended to make sure that prosodic detail is rendered appropriately. Although SSML includes a fairly rich set of elements, here the overall concept of *how* to render style and expression is rather poor, resulting in failure to manage detail which is not really part of what has been termed *neutral* prosody (see Chapter 30 for an explanation of this point).
3 *Others*. One or two elements have been added for completeness, but they are not significant in having an effect on the phonetic rendering of the input content of documents.

As with any XML-based markup, elements have their attributes, some of which may also appear under different elements. We list here under each main element its principal attributes together with their possible values.

18.5.1 Document Structure, Text Processing and Pronunciation

- **<speak>** This is the root element for any SSML document.

  - *Attribute:* **xml:lang** qualifies the element by specifying the language to be used. This attribute is also defined for the elements <paragraph> and <sentence>.

- **<voice>** specifies the voice to be used by the synthesis processor, including the language if it has not already been specified globally as an attribute of <speak>. The point here is to enable local change of language if required.

  - *Attribute:* **variant** specifies a preferred variant of the voice–an alternative name.

- **<paragraph>** and **<sentence>** are elements used to mark the internal structure of input texts. These elements are optional, but if they are not used it is down to the synthesiser processor to determine the document's structure. These elements contain text and sub-elements needed for further rendering details.

- **<say-as>** is used mainly for text normalisation–that is, expanding abbreviations, dates, numerals etc.

  - *Attribute:* **type**–for example: *pronunciation* type ["acronym", "spell-out"]; *numerical* type ["number", "ordinal", "cardinal", "digits"]; *time* and *measure* types ["date", "time", "currency", "measure", . . .]
  - *Further attributes:* **interpret-as**, **format** and **detail**. The application of these attributes is complex, but leads eventually to greater control of text or symbols which synthesisers might find difficult or lead to variable or unpredictable results–the idea is to increase control and therefore versatility (the original W3 specification should be consulted on these details).

- **<phoneme>** is intended to enable explicit characterisation of pronunciation, bypassing pronunciation processes in the attached text-to-speech system.[1]

  - *Attribute:* **alphabet** specifies which phonetic alphabet; for example: ["IPA", "SAMPA", etc.].[2]

- **<sub>** replaces the contained text in the original document by the text specified in the element's alias attribute.

  - *Attribute:* **alias**–substitute the following quoted text.

18.5.2 Prosody and Style

- **<prosody>** enables prosody control while <prosody> is open (that is, before </prosody>).

  - *Attribute:* **pitch** specifies the fundamental frequency baseline in hertz (Hz), a relative change, or the values ["default", "low", "medium", "high"]; how the baseline *concept* is interpreted by different synthesisers may vary.[3]
  - *Attribute:* **contour** is used to specify a fundamental frequency contour. Note that this is explicitly *not* an intonation contour. An intonation contour is an abstract object either within production phonology or perception. The contour itself is defined by specifying points in the text at white space marks (that is, at points usually but not always between words) followed by the fundamental frequency in Hz. The marker points are indicated as percentages of the contained text duration from its start. Thus, for example: <prosody contour = "(0%, +15Hz) (5%, +25Hz) (10%, +35Hz) (25%, +25Hz) (50%, +15Hz) (100%, +5Hz)">.

---

[1] There is a potential problem here because SSML is unclear whether the intention is to specify a phonological or a phonetic rendering. A phonetic rendering could be addressed to the final stages of the synthesis processor, whereas a phonological rendering would require further processor rendering at the phonetic level–equivalent, for example, to coarticulation.

[2] The International Phonetic Alphabet is recognised by SSML as embodied in a subset of unicode symbols.

[3] There is a potential problem here because the word 'pitch' is normally served to mean *perceived fundamental frequency*, whereas in SSML it is used to mean *actual fundamental frequency*. The relationship between the two is not linear.

- *Attribute:* **range** specifies the fundamental frequency range in Hz, a relative change, or the values ["default", "low", "medium", "high"].[4]
- *Attribute:* **rate** specifies the rate of utterance delivery in words/minute, a relative change, or the values ["default", "slow", "medium", "fast"].
- *Attribute:* **duration** specifies the duration of the specified object or item in seconds or milliseconds–but consistently one or the other.
- *Attribute:* **volume** specifies the amplitude for the final audio using a range of 0.0 to 100.0, a relative change, or the values ["default", "silent", "soft", "medium", "loud"].[5]

- **<emphasis>** means synthesise the wrapped text with emphasis. Note that *emphasis* is not made explicit in a way which corresponds to any particular usage in linguistics. We assume is means 'contrastive emphasis' or 'prominence', or perhaps 'focus'.

  - *Attribute:* **level** specifies the degree of emphasis using the values ["none", "reduced"]– appropriate here because researchers are uncertain as to the acoustic parameters of emphasis (because they vary), and because emphasis may be handled differently by different synthesis processors owing to the theoretical uncertainty.

- **<break>** is formally an empty element for inserting a pause of specified duration. It is used for controlling pausing and marking prosodic some boundaries.

  - *Attribute:* **size** expresses the *length* of the pause expressed in *perceptual* terms as marking a perceived boundary of some significance. Values are ["none", "small", "medium", "large"]. Its use is optional.
  - *Attribute:* **time** specifies the *duration* of the pause expressed in *objective* terms. Values are in seconds or milliseconds. Its use is optional.

18.5.3 Other Elements

- **<audio>** triggers an embedded audio file.

  - *Attribute:* **src** points to the name of the audio file.

- **<mark>** inserts a marker for internal referencing purposes within the current SSML document, or for external referencing from other documents. When the speech processor detects the <mark> element it is directed to issue an event with the name of the marker.

  - *Attribute:* **name** specifies a string representing the event's name when <mark> is reached. It is *required*.

18.5.4 Comment

One of the problems associated with some of the elements and their associated attributes is the way in which physical and cognitive terms have been mixed. We have indicated when this occurs. Sometimes is it obvious which is meant, but at other times it may be confusing

---

[4] Instead of "low" and "high", we feel that "narrow" and "wide" respectively would have been more appropriate.
[5] The terms *soft* and *loud* are normally used to describe *perceived* amplitude and are unhelpful here. We feel that "low" and "high" of objective intensity or amplitude would be more appropriate.

both for the human reader of the document (remember one criterion for SSML documents is that human beings can read them) and for the synthesiser processor. However, we are not saying that perceptually based terms should not be used. They are entirely appropriate where the acoustic parameters which characterise the element may not be clear (as in the case of perceived stress or emphasis), or where–perhaps because of this uncertainty–the way the property is rendered by the synthesis processor may be unknown, variable or otherwise unpredictable. This is one of the areas where the rendering process and its results may be different for different synthesisers.

19

SABLE

SABLE (SABLE Consortium 1998) is a markup language focussed on using synthesis for distributing information–speaking information from databases is the dedicated goal of the synthesis environment. Such inputs are often characterised by a predominance of special formatting or symbols. That is, special extra-textual properties such as spreadsheet or database layout, and symbol definitions or expansions such as numbers, dates etc., take on a high priority in the markup because they are often difficult or even impossible to derive by rule.

Eventually SABLE is seen as playing a significant role in assisting speech synthesis rendering in generalised multimedia systems, so that tags (elements) are included which can directly address media in general–matching up with tags appropriate to dedicated media, such as video or audio data streams. Employed in this way, SABLE has the potential for assisting in the control of integrated media systems.

We shall not go into details of SABLE here, but the specification provides explicitly for three types of tagging in the markup:

1 *Speaker directive tags*. These elements of the markup enable explicit control of emphasis, break, pitch, rate, volume, pronunciation, language and speaker type.
2 *Text description tags*. These are elements for identifying names, dates, telephone numbers and other special symbols, as well as rows and columns in a matrix or table, etc.
3 *Media tags*. These are designed to facilitate linking and synchronising media other than speech (e.g. video).

In general, the speaker directive tags have the potential for covering many of the features found in SSML. The text description tags explicitly extend what is to be found in SSML, and the media tags generalise beyond the intentions of SSML.

Developments in Speech Synthesis   Mark Tatham and Katherine Morton
© 2005 John Wiley & Sons, Ltd.   ISBN: 0-470-85538-X

20

The Need for Prosodic Markup

20.1 What is Prosody?

Prosody comprises the features of an utterance which span its individual sound segments.
It is the basis for expression in speech. The 'suprasegmental' features of prosody span the
segments of speech, and are usually rendered in the final soundwave as changes of fundamental frequency (the f0 contour), patterning of segmental durations and pausing, and
patterning of amplitudes. Such acoustic features constitute instantiations of more abstract
underlying phonological features such as intonation, rhythm and stress. The relationship between
the phonological and acoustic features is complex and nonlinear.

20.2 Incorporating Prosodic Markup

In Part IV we discussed the fact that plain text does little to encode speaker prosody, with
only the basic punctuation to assist the assignment of prosody. One way around this is to
extend normal representations to include an explicit encoding of an appropriate prosody. In
the early days of text-to-speech synthesis the focus of research effort lay in retaining a plain
text input and avoiding any extra marking of the text–on the grounds that this usually required
expensive or unavailable human intervention. With increased focus on naturalness it has been
realised that more progress is to be made by relaxing the earlier requirement of a plain text
input and instead allowing markup of features, such as prosody, which are proving relatively
elusive in simple text-to-speech systems. Paralleled with this change of focus is the increasing requirement for niche (and therefore more manageable since they are more predictable)
applications, such as read-aloud email. The usual way of regarding prosodic markup is as
some kind of intermediate representation between the plain orthography and *how* it is to be
spoken. We shall see that this is rather too simple an approach, and that just spotting email
headers and dates, for example, undermines what is potentially the key to much more
natural synthetic speech.

  Orthography is a symbolic representation of the morphemic structure or, when suitably
interpreted, forms the basis of the phonological structure of an utterance–so that embedded
prosodic information should also be symbolic at the same level of abstraction. Suppose, for
example, that a couple of paragraphs of text in English are to be spoken. Initially the representation is normal orthography which is soon converted to an underlying phonological
representation. In most synthesis systems prosodic markup can occur before or after the
conversion. Just as the phonological representation needs processing to more suitably form

Developments in Speech Synthesis   Mark Tatham and Katherine Morton
© 2005 John Wiley & Sons, Ltd.   ISBN: 0-470-85538-X

the basis of a phonetic representation, so the prosodic representation will need similar processing.

20.3 How Markup Works

Generally the use of markup to supplement plain text input is regarded as a kind of annotation of the text–the addition of extra information which will eventually result in an improved synthetic output. We believe that this is not the best way of approaching the problem. We prefer to think of prosody as forming the basic structure within which the segmental detail of utterances is deployed. We have discussed this approach in terms of a hierarchy of wrappers (Tatham and Morton 2003), which characterise data structures appropriate for encapsulating the segmental aspects of utterances. This approach enables greater and more insightful generalisation since the wrapper structure exists independently of any one utterance, and enables comparison *between* utterances in terms of their place in the data structure hierarchy. Thus two or more utterances can be seen to communicate the same emotion *via* the shared prosody which wraps them (a parallel comparison), or sequential utterances can be seen to move slowly between emotions (a sequential or serial comparison) as wrappers progress within a hierarchy.

The mechanism which enables generalisation is the *schema*, equivalent formally to the abstract general case. In linguistics, for example, the general case for, say, the syntax or the phonology of a language lies in a transformational grammar characterisation. (See Chomsky (1957) for the first reference to transformational generative grammar.) This is a characterisation which holds for all sentences in the language, and in that sense characterises the language itself; it is not a characterisation which applies just to one sentence. Such instantiations of the general case are given in transformational grammar only as exemplar derivations and do not form part of the grammar. Similarly the markup of a particular document for input to a speech synthesis system would properly be an instantiation of a general case statement which characterises all that is possible in terms of the markup, but without reference to this particular document. The general case statement is the schema. This is also the mechanism for checking the legitimacy of a markup–the particular markup is validated against the schema, just as in linguistics a particular sentence is validated against the grammar. In both areas the focus is on legal data structures and how they interrelate.

20.4 Distinguishing Layout from Content

Unconstrained, XML does not of itself distinguish between layout and content. Careful choice and hierarchical arrangement of elements and their attributes is essential. But this obviously cannot be achieved if we ourselves are unclear as to the distinction. An XML-based markup like HTML is clearly mostly about layout–in this case screen layout of text. And we may want some of our prosodic markup to deal with layout in the sense that it can be used, say, to identify email headers, date stamps, routing information etc. But we have to resolve here an ambiguity. Routing information appears by convention, for example, in a particular place in an email and takes a particular format–this is layout information, but the detail of the routing information is clearly content. A simple question to ask is: Does the identification of layout immediately lead to identification of content? We need to get away from the older association with layout and turn more in the direction of marking content–preserving both but being absolutely unambiguous as to which is which, perhaps by different types of element. So for example in SSML the elements <paragraph> and <sentence> are in one sense layout-oriented in that they contribute toward *identifying* the document structure, but on the other hand are content-oriented in that they contribute toward *determining*, say, intonational phrasing within sections of the document. On the other hand, an SSML attribute like pitch (on the element <prosody>) is clearly a content marker, whereas the element <audio> is clearly a layout marker. There would be a clear gain from unambiguously sorting out which elements and which attributes can relate unambiguously to layout or content.

20.5 Uses of Markup

The most effective use of markup at the moment lies in those areas where the input text does not contain or cannot be re-rendered to contain vital information; and among these prosody is the most important. But before coming to prosody there are a few other areas where markup improves output by making up for inadequacies in the synthesis processor. Among these are areas which may have little to do with the content of the input text, but perhaps more to do with the social context in which it is to be spoken:

- choose language
- define accent within language
- change voice (i.e. move to a different speaker).

Suitable markup elements can be devised to enable these choices with appropriate nesting of the different elements if called for. Thus, using a generic markup language, we can imagine a young woman from London speaking to a young man from Los Angeles:

```
<synthesise xml:lang="British-English" accent="Estuary">
        <voice variant="young-female">
        When are you coming to London?
        </voice>
</synthesise>
<synthesise xml:lang="American-English" accent="Southern-
California">
        <voice variant="young-male">
        In about two weeks.
        </voice>
</synthesise>
```

A synthesis processor of the unit selection type uses two different databases taken from its stored range. The device would fall back to suitable defaults if exact-match databases are not available.

But within the document there may be content which the synthesis system cannot be expected to cope with.

- The pronunciation of some proper names may not be able to be deduced by rule and may be unusual enough not to be in a fall-back pronunciation dictionary or appropriate database.

- Foreign words may have to be 'regionalised'–say, in the case of an original French word where a listener would expect a kind of hybrid pronunciation (for example, envelope ['ɔnvəloup], or restaurant ['restrɔnt]).
- Non-speech sounds are produced by speakers, including hesitation phenomena and other paralinguistic sounds (for example, *er . . .* or *hmm!*).
- Some elements direct content flow, in the sense that they may call in text from other documents or branch to other media, or indeed call simultaneous processes in other media (e.g. setting up background music for the spoken document or accompanying light effects and video).

20.6 Basic Control of Prosody

Prosody is the main area where careful use of markup will improve the synthetic output, contributing enormously to perceived naturalness in the speech. There are several models of prosody which can underpin the markup, but basically all models approach the problem by identifying features of the structure of the text and then manipulating details, in particular those which signal prominence of some kind, within the structure. The word 'text' (as always) means text to be spoken because its potential prosodic structure does not necessarily correspond to its non-speech properties. An essential consideration here is that text to be spoken has to be text that is *suitable* for speaking–a truism, but something which is frequently overlooked as synthesisers (and even human beings!) struggle to speak text that just does not lend itself to being spoken. It has to be remembered that not every text author is gifted in writing the way people actually speak.

A generic structure behind such models of prosody is a hierarchical arrangement of topics, paragraphs and sentences:

```
<topic>
     <paragraph>
           <sentence/>
           <sentence/>
           . . .
     </paragraph>
     . . .
</topic>
```

where a topic can contain a number of paragraphs each of which can contain a number of sentences; a document, of course, could contain a number of topics. Sentences are sub-divided in some prosodic models into different types of phrase. There can be more than one phrase within a sentence domain.

As always, synthesis processors will have their own way of rendering such high-level structural elements; but in principle all systems have to signal their boundaries if perception is to make sense of utterance structure. There are a variety of ways of doing this, but compatible models of human speech production usually refer to phenomena like the duration of pauses, variations in speech rate or changes in the range of fundamental frequency; these are physical phenomena which can be referred to quantitatively. Phrase boundaries are rendered partly by pauses of shorter duration and by changes of fundamental frequency contour within the ranges and contours established for the sentences within which the pauses occur. At a more abstract and potentially more ambiguous level, these models often refer to 'boundary tones' as potential delimiters of the effective domain of structure-focussed elements. A boundary tone is usually taken to mean some kind of contour modification at or signalling a syntactic break, like a phrase boundary.

Within the prosodic structure we find *prominence* or salience, an abstract quality roughly correlating acoustically with fundamental frequency contour disturbances and segment durations greater than normal. There are a number of important point to be made here.

- Prominence is an abstract term associated with the reaction of the listener to the utterance being perceived, but it is also part of the speaker's plan to give prominence to some object in the projected utterance.
- The acoustic correlates of prominence are variable, and can correspond to increased amplitude, an unexpected change in fundamental frequency and an unexpected increase in segment duration–or any unusual mix of all three parameters. Sometimes these three parameters are grouped under the unhelpful assertion of 'increased articulatory effort'.
- The segment which forms the basis of prosody is the syllable–an ambiguous unit since on the one hand it is an abstract unit of planning and on the other a physical unit of rendering. The correlation between abstract and physical syllables is nonlinear and sometimes hard to arrive at.
- Prominence means that the acoustic signal and what is perceived are somehow unusual, implying departure from some norm. Characterising this norm has proved difficult enough to warrant questioning the basis of the model (Tatham and Morton 2003).

We are not concerned with where the speaker's need to introduce prominence comes from, but generally there are underlying semantic and pragmatic considerations of importance or emphasis being channelled into the planning process (see Part VII where a possible dynamic production model for human beings is explained).

It cannot be over-emphasised that, despite considerable effort on the part of researchers into human speech and engineers concerned with speech technology, there are still two wide areas of weakness in our understanding of how prosody works in human speakers:

1 The characterisation of prosody in terms of an abstract structure with the potential for expressing nuances of semantics, pragmatics and expression in general is still fragmented– the *underlying static model* suffers from a poor theoretical foundation.
2 The nonlinearity of the correlation between cognitively planned dynamic utterances based on the possibilities enumerated in the static model and the final soundwave of instantiated utterances remains largely unfathomed.

These problems lead, in the synthesis model, to two corresponding areas of difficulty:

1 The provision of high-level markup is inevitably inconsistent and problematical, particularly if the markup is to be performed automatically.
2 The reliance on different synthesiser platforms to interpret and then render the markup inevitably falters. This is one of the areas of synthesis where implementations vary considerably, making for poor inter-platform robustness.

As examples we might consider

- the disagreements among theorists as to how to determine the prosodic structure of utterances, especially the extent to which syntax is involved;
- the poor agreement on the acoustic correlates of any kind of expressive content.

It would be a serious mistake to think that devising an annotation or markup system for the text in our synthesis documents solves the problem of prosody. This does not mean, however, that we should not press on. Hopefully development of the theory of prosody in human beings will improve in the not too distant future, and meanwhile much practical experience will have been gained in improving markup procedures. We discuss elsewhere how markup has to be improved, but let us take a single example which is very relevant here: expression in speech is a *continuously varying* phenomenon. Notice, though, that markup such as we have at present cannot handle this simple fact:

```
<utterance>
        <forcefully> I was so indifferent to him at the beginning,
        </forcefully>
        <expectantly> but I soon realised that </expectantly>
        <happy> I could easily spend the rest of my life with him. </happy>
</utterance>
```

In this example we are marking at an abstract level a rapidly changing emotion which corresponds roughly to the words being spoken–many do not. The markup is clumsy if it is trying to capture how one emotional state gives way to another. To a certain extent we can make up for the abrupt shifts between emotional states by introducing a rule of rendering which allows for blending of the emotions as the sentence unfolds as an actual soundwave. This would be analogous to characterising coarticulation in speech production theory as a blending of targets as time unfolds, but it runs into exactly the same problem associated with correlating abstract notional time (object sequencing) with gradually unfolding events in the physical instantiation–the unfolding is nonlinear and varies between instantiations. This either precludes introducing a rule for converting between sequencing and unfolding or makes it very difficult to think one up which will work in a variable way.

20.7 Intrinsic and Extrinsic Structure and Salience

Text leading to utterances can be modelled as having *intrinsic* structure and salience. These are the structure and prominence features which derive from the way language, *the* language, and the relevant semantics and syntax pan out, all things being equal. So, for example, there is a structure *intrinsic* to any utterance in any language, and to any utterance in English, and to the way certain features of structure and prominence are handled for any utterance.

Whatever these structural and prominence features are, they underlie any logically subsequent optional *extrinsic* features, which might be said to expand on the situation where all things were equal. This simplistic model is quite common in approaching several areas in linguistics; for example, the hierarchical relationship between different types of allophone–variants of speech segments grouped within sets and deriving from underlying invariant, but more abstract, segments.

> In English we might here be reminded of the way in which speakers and listeners can be made aware that there are two different *extrinsic* allophonic segments to be derived in strictly constrained environments from an underlying abstract /L/ object. One type of /l/ is used in syllable onsets and the other in syllable codas. But subsequently each of these two is rendered as a potentially infinite number of variants by properties intrinsic to the speech production mechanism–intrinsic allophones outside the perceptual sensitivity of native speakers. Analogies are never perfect, but the inbuilt occurrence of intrinsic allophones reflects the way allophones are, just as the intrinsic features of the structure of utterances are equally inevitable. The optionality of extrinsic allophones parallels the optionality of extrinsic structural and prominence features of the utterance.

In this sense we are free to vary some aspects of structure and prominence in text to be spoken, whilst being tied to other fixed aspects. We would like to see the markup reflecting this, so that it becomes transparent to the human reader of marked up texts, and to the synthesiser which must render the texts, just *which* markup is basically optional and which is *not*. The object here is not to enable the synthesiser to change the intention of the markup, but to enable synthesiser designers to spot what aspects of structure and prominence *must* be implemented in the synthesiser's rendering algorithms, and which are open to some leeway in interpretation.

One useful consequence of nesting extrinsic markup within intrinsic markup would be identifying strings of pragmatically linked sentences, as opposed to those which may be equally sequenced in terms of layout, but are less linked semantically. Intonation, for example, will be different across sentence boundaries in the different cases. Perhaps less frequently, but therefore more dramatically, would be the signalling of linked paragraphs within the sequence of paragraphs in a document. Such a device would assist the problem mentioned earlier of continuously variable prosody seen as a feature of expressive content in speech.

The question of intrinsic *vs* extrinsic markup of the structure and prosody of text hinges on whether markup is being used to bring out what might have been in the text if a better orthography had been used–that is, can it reveal the writer's intentions as they would have appeared if the writer had personally spoken the text. The writer has control over the semantics and syntax of the text etc., including the choice of words and a few other marks. But the writer does not have control over the extrinsic prosody–this is assigned by the reader (or the markup engineer, or the synthesis system). The task is to assign extrinsic markup of these properties as though they were being optionally assigned by the writer–a necessarily impossible task since we cannot get into the mind of the writer and the orthographic system does not allow the writer to tell us. However, the intrinsic markup will be the same whether the writer had been allowed to use it or whether we later apply it to the text. The intrinsic markup is shared between writer and reader–the extrinsic markup would ideally be shared, but cannot be. Put another way: intrinsic markup is redundant (because it is predictable), but extrinsic markup is critically *not* entirely predictable.

20.8 Automatic Markup to Enhance Orthography: Interoperability with the Synthesiser

The problem with automatic markup is trying to work out what the original writer intended. We have his or her words and punctuation, and can have a reasonable shot at a basic pronunciation, but we do not have access to the writer's intentions as to accent and prosody beyond what is intrinsic to the language. Thus we can correctly assign, for example, falling intonation toward the end of a typical sentence, but do not know whether the writer might have preferred a sentence to be spoken with a rising querying intonation at the end. This is not a problem, of course, if the writer can personally assign markup as a kind of enhanced orthography, or proof-read the suggested automatic assignment. This checking-back approach, however, is unlikely to be the majority usage, and certainly cannot be used for 'on-the-fly' rendering in variable circumstances or environments.

The purpose of this additional markup, though, may be to add extrinsically sourced information; the task may be either to interpret someone's intentions or to enhance those intentions. An additional extrinsic function may be to provide alternatives where the writer did not explicitly intend them–for example, the text could be marked to be read in alternative accents on demand.

The possibility of alternative interpretations of markup introduces the concept of an intelligent interpreter. Markup systems for any purpose rely on a browser (in our case this is the synthesis processor) able to interpret the markup. But there is also the possibility that the browser can intelligently select from a permitted range of alternatives. The markup would specify what is available, the interpreter would chose a possibility appropriate for the task in hand. We can imagine an interpreter with more than one input, and which incorporated an *intelligent agent* to select the right markup dependent on information from another input channel.

As an example we could consider a dialogue system whose purpose it was to speak material from a database for a human user. A recognition device could have detected that the user has special needs, and that a particular way of speaking is necessary. 'Way of speaking' is analogous to using a particular expressive tone of voice which might be selected from a range which was pre-determined by the text's markup. The markup could specify that under certain circumstances the speech synthesis can use certain possibilities from a particular range, but under others must exclude certain possibilities. The intelligent interpreting agent would make an appropriate reasoned selection depending on

- text content itself
- the range of possible ways of speaking it
- additional environmental information being supplied by the recogniser
- the range of prohibited possibilities in this environment.

It follows from the idea of an intelligent interpreter that the basic assumption made by synthesis markup systems that the speech synthesis will use its own method of dealing with the markup needs reconsidering. The original idea was that elements in the markup would be interpreted in a platform-dependent way, while preserving the platform independence of the markup itself. This makes for little or no *active* inter-operability between the markup and the interpreting platform. It is clear to us, though, that eventually provision must be made for not only such inter-operability but also for the interpreting synthesis platform to introduce a measure of independence into how it interprets markup instructions on any one occasion. For example, the weightings on particular values may change during a dialogue– the need for faster or slower delivery rate of the computer speech may arise.

The concept is very simple, but implementing it will not be a straightforward matter. Once achieved, though, this approach will produce not only a more natural sounding speech, but a more appropriate speech rendering. Naturalness and appropriateness are not the same thing, and it clearly is not possible for all aspects of either naturalness or appropriateness to be contained in the original document markup, if only because they will be dependent on unpredictable environmental considerations at the time of speaking.

As we introduce the concept of *intelligent* interpreting of the synthesis document we move way beyond the original idea that the document presents a once-for-all way of speaking its content. The provision of dynamic markup and variable, intelligently directed, interpretation of that markup is something we shall have to face in the not too distant future. The reason for this is that perceived naturalness in speech goes beyond appropriateness for the text and extends into appropriateness for the environment and occasion–two constraints which are relatively unpredictable. The purpose of markup shifts from being prescriptive to being *assistive* with respect to subsequent rendering. A useful analogy can be found in the way drama scripts are interpreted by actors: a contemporary or modern-setting production of Shakespeare, for example, would not be rendered in the same way as one attempting to be authentic to the point of reproducing the way sixteenth century plays were staged. We assume the playwright would be neither surprised by nor opposed to using his plain text and markup (stage directions) in an appropriate original fashion. To push the analogy further, it is the interpretation which brings life to the play in a fresh way which cannot be foreseen by the script. We can see that at least some of that 'life' lies in its novelty; in the same way, 'naturalness' of synthesised speech will ultimately include unpredicted novelty–we assume the listeners are looking to find precisely this in what they hear.

Complete novelty by definition precludes exhaustive markup. Markup needs to be there as a constraint on novelty, just as the orthographic text needs to be there to prevent the synthesiser speaking randomly. We could even advance the hypothesis that prosody is not included in orthography because we expect it to have an unpredictable element–something we do not expect of the words themselves. Contrary to what might be assumed to be the ultimate conclusion of the W3 effort in specifying, say, SSML, we do not want to specify in all aspects exactly how the synthesiser is to render our document–paradoxically that would actually prevent a level of naturalness we might want to achieve.

This does not mean that there will not be occasions on which we really do want to specify everything. The analogy here is with a legal document. Lawyers and attorneys use special language to exclude as much ambiguity and alternative interpretation as possible. Maybe in the same way some documents to be spoken really do need as much constraint as possible on how they are to be spoken; alternative renderings need to be prevented.

20.9 Hierarchical Application of Markup

Systems of markup like SSML and SABLE adopt a convention like that of XML partly because it is hierarchical in nature, permitting the nesting of elements within elements. Additionally XML will enable variable attributes to be attached to individual or multiple elements. What is not usually discussed, though, is the hierarchical *application* of the markup.

The textual content within a document (i.e. the words that are to be spoken) is hierarchically processed in two stages: high-level and low-level synthesis. And *within* each of these stages the processing is also hierarchical. A cascade of processes from input text through to a file of instructions to a formant synthesiser or a unit selection concatenative system successively rewrites the input until the results conform to the output file specifications. This is generally how all synthesis systems proceed, though there may be general or local details which are not actually hierarchically organised. The markup systems devised to augment the plain orthography of a document's text are, however, applied to that text at a single level–the input level to the entire system. Thus unmarked text is directed away from direct input to the speech synthesis, rewritten with added markup and then redirected back to the speech synthesis input. The markup is therefore applied 'flat' to the text as a single process: one of intervention or supplementation.

But clearly markup is useful at different levels in the hierarchical synthesis process itself. Some markup, for example, may be useful in interpreting textual conventions not obviously present in normal orthography, other markup for added detail to the exact pronunciation of an individual sound, and yet other markup for generating details of features of the acoustic signal such as fundamental frequency fluctuations.

That markup is applied flat in a single process, yet applies to different stages in subsequent processing explains why most markup systems seem to mix different element types indiscriminately; for example, abstract elements like <emphasis> are interspersed with more concrete elements like <phoneme> in SSML. An element like <prosody> should be applied much earlier than <phoneme>, and so on. As a general rule phonology-focussed elements should be applied before phonetics-focussed elements. The elements themselves do not include the information as to where in the synthesis process hierarchy they should be applied, and in any case this potentially differs from system to system. We feel that it is worth considering whether automatic markup might be better applied at different stages in the synthesis process, rather than all at the start in the input document itself. As a very minimum it would be theoretically sound to apply phonological elements (logically) earlier than phonetic elements.

20.10 Markup and Perception

One problem facing synthesiser designers is to what extent perception should be taken into account in generating the soundwave (Liu and Kewley-Port 2004). It is well known that speakers adjust their pronunciation depending on a number of different perceptual features (Tatham 1986a,b; Lindblom 1990), and some of the more sophisticated production models include a production agent whose job it is to continuously predict the perceptual results of hypothesised speech output. These models account for the observed fact that speakers continuously vary, for example, the precision of their articulations and hence the conformity of the soundwave to predetermined norms. For many speakers the overriding consideration seems to be that speech is to be articulated with as little precision as possible consistent with adequate perception. If this is the case then all current markup proposals will produce too consistent an output since they fail explicitly to consider the perceptual effects of the results of applying markup. Put more simply: the perceptual environment has a bearing on how speech is to be pronounced; and moreover perceptual environment is continuously varying.

The perceptual environment includes a number of variables including simple considerations like the amount of background noise–greater noise needs a louder signal but also a signal pronounced more slowly and more carefully (i.e. with greater precision). Human beings adjust their speech automatically to cater for the noise environment. But there are linguistic factors too: if the speech is semantically, pragmatically, syntactically or phonologically predictable then the precision of articulation can be reduced, and vice versa. Listeners expect this to happen, and if it does not they are likely to judge the speech to be less than natural. There are two difficulties here in modelling speech production for synthesis:

- The data as to the required accuracy comes from a predictive model of the perception of speech *within* the production system.
- Markup of the XML type is not at all suited to continuously variable phenomena, because it does not 'span' objects in an overtly variable way, and because it does not make explicit provision for data channels other than the text itself.

An XML element is opened and later closed, applying equally to all text within its domain. We have already encountered a difficulty with this constraint when it comes to continuously variable expression in speech. It is a problem also when we need to keep switching precision of articulation, and hence of acoustic signal. But it is also hard to introduce another data source into the document, particularly one which has not been predetermined but which is on-going *during* the speech production. How do we adjust the effects of markup in temporal correlation with changing perceptual circumstances? How do we make markup conditional on something which has not yet happened?

The last question is relevant for good synthesis. We can observe a very important feature of human speech production–speakers adjust the precision, intonation and rate of delivery of speech in accordance with *what we think the listener will make of it*. We continuously change our speech based on a prediction of how it *will be* perceived.

The solution we have introduced in our current model of speech production (Tatham and Morton 2004) was to introduce a continuously running predictive model of speech perception which could test the perceptual results of planned utterances *before* the plans were finalised and passed for rendering. It is not difficult to build this idea into speech synthesis strategies, of course, but it is hard to characterise it in a traditional markup system like XML and therefore SSML and other similar systems.

We do not have a solution for this at the moment, but are simply bringing it up because variability of this kind is a key element in the humanness of speech, and therefore becomes an important focus for strategies for introducing more naturalness into synthesis. (See Chapters 21 and 22 for a discussion of the model of human speech production.) It may be possible to adapt existing markup strategies to take this into account, or at least to make provision for introduction of the idea at some later stage.

20.11 Markup: the Way Ahead?

To a certain extent markup is clearly the way ahead for text-to-speech synthesis. The reason for this is that it allows an explicit means of annotating text with information it does not already contain. Text represented as ordinary orthography is a fairly good characterisation of the syntactic and lexical (and therefore morphemic) structure of the text. It is a poor representation of the semantics and pragmatics, partly because it is impoverished with respect to prosodics. The latter cannot be a sustainable criticism, though, since text is not primarily intended either as a record of something that someone has spoken or as a script for reading out aloud. Spoken language always has prosodics which are used to convey structure, general pragmatics and expression, as well as emotive content. Written language can convey the same features, but not through the medium of prosodics–there cannot be a prosodics of written language.

Linguists recognise that there are spoken and written versions of languages, and have a good grasp of the differences. Problems arise, though, when we try to switch modalities–write down speech (not a serious problem), speak out written text (a very serious problem). Efforts to add spoken features to written text are poor, as any researcher or user of text-to-speech systems knows. Yet they *can* be added with a great deal of success–by human beings reading text out loud. The results are a little strange when compared with genuine speech, but certainly very, very much better than our attempts with synthesis. So, all we need to do is simulate what people do when they read aloud!

From the point of view of the segmental pronunciation of written text the problem is not great. Ordinary orthography can be rewritten to give an underlying abstract representation of how it might eventually be spoken. And where this conversion cannot be done by rule it is a simple matter to use lookup tables (in the general case) or markup (in more specific cases). SSML's <say-as> and <phoneme> elements are examples of markup which can take care of most of this problem. The result is a phonological re-representation of the text. It must be understood, though, that not all the necessary detail is represented for simple conversion to a soundwave. This is the main reason why there is still progress to be made in making the segmental aspects of speech better. This is mostly true of formant synthesis and concatenative systems which use small units, but it is also true of variable-length unit concatenative systems because even very large databases cannot capture all pronunciation possibilities.

But although conversion of orthographic representations to phonological representations can be done automatically for a high percentage of any one text, the same is not true of prosodics and what prosodics is used for. The reason for the difference is that for most languages even a non-phonological orthography as used in, say, English or French historically was intended to represent how the language was spoken–so there is a vestigial phonology present which can be augmented quickly by rule or some other automatic means. But there is really no prosodic information present at all. It is often claimed that punctuation indicates a rough prosodic structure, but this is an exaggeration. Punctuation indicates the structure of the text, and this is syntactically based (sentences delimited by full-stops and broken up by commas) and semantically based (the addition of paragraph and indentation or bullet conventions etc.). These can be interpreted as helpers for prosodics only by coincidence, since prosodics also needs structural conventions. Punctuation conventions were unlikely to have been introduced into writing for the purpose of indicating spoken prosodics; whereas at least in western languages certainly spelling *was* used historically to indicate how words etc. sounded.

This is where markup comes in. It can be used to augment ordinary orthography to indicate prosody. And since we are starting from scratch we can make it explicit and consistent–unlike the historically evolved representation of segmental pronunciation. So, armed with a knowledge of spelling and a suitable markup system, an author can write text that can be spoken aloud complete with its intended prosody. We shall see later that there is much, much more to it than this.

But this is not the immediate problem we face. Researchers in speech synthesis want to add the prosody without the help of the original author. A fair measure of success is possible for conversion of spelling to a phonological representation, but attempts to augment spelling for prosody have met with almost no really acceptable success–for the reasons already discussed. The problem remains because there is a real need for us to continue to seek ways of speaking out old text or new text where we cannot count on author annotation of prosody.

20.12 Mark What and How?

The task is to add prosodic markup to plain unmarked text. This is effectively a simulation of what a human reader does in formulating a plan for an utterance based on seeing printed text. The process is not entirely understood, but there may be certain characteristics of the human process which we can observe and try to emulate.

The process is analogous perhaps to what is involved in deciding how to pronounce the text segmentally–derive from the orthographic representation a suitable underlying phonological representation and proceed as though this were an original phonological representation the readers would have developed for speech which *they* originate.

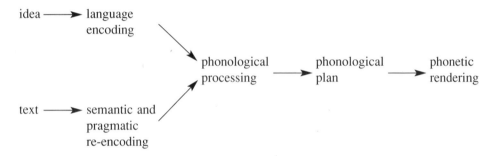

For original speech the meaning is already present in the underlying *idea* which is to be encoded as language. The processes involved are those necessary to following the agreed encoding conventions that we call language. Once the idea has been formulated as a sentence it can be phonologically processed to derive a plan for speaking. The plan contains all the phonological (including prosodic) information necessary for phonetic rendering to proceed.

With text the original meaning is not present. It has to be interpreted from the text by the reader. To some extent the way the author has done the orthographic encoding helps. But the interpreting process is a very complex active one which involves the reader bringing something personally to the task. The reader's knowledge of the language is only part of the task; the remainder depends on the reader's experience of the world far beyond the scope of the structure of the encoding language.

Once we understand the difference between original speech and reading text aloud by the human being, we can ask the question: Could we build a device that could bring human-like experience to the task in any realistic way? There is no need to debate this: the answer is 'no' or at best 'not yet'. True, we might be able to imagine toy or very restricted domain systems which could have a shot–but their 'success' at the task would be heavily dependent on the tiny domain, *not* on their ability to perform the task in general. Some researchers believe this is the place to start, because at least we can begin to understand some of the principles involved, and these might form the basis of an ever-expanding domain.

The current demand, though, far outstrips our ability to build a device which can, in effect, 'understand' text so as to mark it up for the prosody needed to speak it. We must therefore back off from the requirement for a fully automated markup designed to augment someone else's original text, which probably wasn't intended to be read aloud anyway!

There is a halfway house we could tap. If text is not generally intended to be read aloud it must be able to communicate something of what it means–the process is not one of the reader adding 100% of the meaning. The less a text was intended to be read out, perhaps the more information it contains. If researchers in artificial intelligence could access this information it could form the basis of a system of augmentation–building on what is there–to come up with a reasonable spoken prosody. Researchers in speech synthesis could do well to explore this avenue a little more.

20.12.1 Automatic Annotation of Databases for Limited Domain Systems

In setting up a system designed to automate the processing of prosodically marking up a database for limited domain concatenated waveform synthesis, Bulyko and Ostendorf (2002) use a so-called 'bootstrapping approach' in which some small section of the database is carefully marked by hand to enable a set of 'prosodic templates' to be set up. The markup is symbolic in terms of phonological prosody rather than in terms of phonetic prosody which would emphasis the physical nature of the database itself. The resultant prosodic templates, together with a decision tree, enable the rest of the database to be marked automatically; that is, their small-scale hand-marking leads ultimately to a set of hypotheses as to the prosodic content of the remainder of the database.

Undoubtedly the success of the system is due to the use of symbolic markup which neatly sidesteps the problems associated with handling variants in the actual signal. Using symbolic markup is legitimised provided the focus is on perceptual judgements of quality in the result-ant signal, since perception itself is about the assignment of hypothesised symbolic rep-resentations to the acoustic signal. See also Wightman and Ostendorf (1994) for an earlier approach.

20.12.2 Database Markup with the Minimum of Phonology

Black and Font Llitjós (2002) raise an interesting question arising from practicalities of database markup of languages which are short on available phonological or phonetic descriptions. There are many languages and accents of languages in the world where the number of speakers is less than a million and, in all to many cases, diminishing fast. Apart from the shortage of time there is also an obvious shortage of trained linguists (and the attendant financial problem) to attempt the task of characterising the phonological structure of these languages, even perhaps to providing the minimum perhaps, usually thought necessary for marking up a synthesis database. The question is whether it is feasible to mark up a database without the luxury of extant phonological descriptions.

Black and Font Llitjós would like to provide what they call 'tools for building synthetic voices in currently unsupported languages', where unsupported means awaiting adequate phonological descriptions–even down to identifying their phoneme sets (their underlying phonological units). Importantly from our own perspective current characterisations (if any) probably also lack adequate characterisations of their *prosody*. The idea is to set up a unit selection system which does not rely on knowing a language's phonemic makeup. Normally unit selection procedures *do* rely on being able to mark the database in terms of phonemes, or for more detail, extrinsic allophones.

There is an assumption that, even though there may be no phonological analysis of the language, at least the language is written and the writing system uses letter symbols. The intention is to use the letter information alone to work out the set of phones for the language; Black and Font Llitjós have already built a number of synthesisers on this principle. In theory, of course, if the letter-to-phone relationship is linear then only a set of simple substitution rules needs to be provided. But these relationships are often nonlinear, and Black and Font Llitjós are aware that the linear substitution idea ('one letter, one sound') is not an entirely satisfactory way of approaching the problem. We would add that it is very important also to distinguish between phonemes, extrinsic allophones and intrinsic allophones (see Part IX) in this context, at least to the point of not confusing levels of abstraction or units within those different levels.

However, despite this their first automated shot is to assume one-to-one relationships between letters and sounds, recognising white space between groups of letters as *syllabic* boundaries–though they do allow the usual expansion of abbreviations, numbers etc. into fully orthographic representations. There is a little confusion in the method as they 'map letters directly to acoustics' and recognise that the more usual technique is to introduce between orthographic symbols and the soundwave what they call an 'intermediate fine phone set'. What is confusing here is the use of the word *phone*, a symbolic representation of putative sounds in the acoustic signal. This confusion, though, is very common in the database markup literature. As we say, it is relatively unimportant, provided levels are not mixed. Normally we would expect strings of underlying phonological segments (phonemes) to be mapped to extrinsic allophone strings which represent symbolically the utterance plan. Individual low-level systems may or may not then have need for a level of intrinsic allophones (symbols *explicitly* representing coarticulated units) or phones (symbols *implicitly* representing coarticulated units).

Advantages of the Black and Font Llitjós proposal include the following:

- Linguistically undocumented or partially documented languages can be synthesised with some degree of success.
- Large quantities of any language could be assembled for database usage or general corpus work in computational linguistics.
- A strategy for markup based on minimal phonological information can be tested on a large scale because the markup is automatic and very fast.

Disadvantages of the proposal include the following:

- It introduces unsound theoretical practices–though we believe that with careful handling the advantages outweigh the theoretical issue since these can be taken care of once the marked databases are developed.
- It depends very much on how closely the orthography matches the pronunciation. If the orthography has been recently developed, the system is more likely to work than if has a long history, during which sound changes have taken place in the language.

20.13 Abstract Versus Physical Prosody

We have seen that there is no prosody equivalent to segmental orthography. Since most researchers, including ourselves, see the need for prosody markup, it makes sense to ensure that prosody notation and phonology notation are in some sense on an equivalent level. This way they can be processed together, with the necessary interaction taking place as and when needed. It is not the case that prosody and segmental phonology are logically other than processes which take place in parallel. In a sense it makes no difference whether prosody markup is applied to the orthography or to the next stage of segmental representation–the characterisation of the underlying phonology, derived either by rule (or some similar means such as a neural network) or by lookup table. We prefer to isolate the orthography and parallel the phonological and prosody markups.

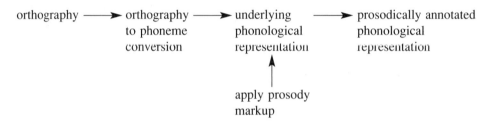

If we apply the prosody markup at this point it must be compatible with the phonological representation; that is, it must be represented at a comparable level of representation or we shall be in danger of violating the consistent level of abstraction constraint on model building. This is the representation from which will be derived the final phonological utterance plan ready for presentation for phonetic rendering. The annotation process occurs at a similar point in the development of an utterance to which it occurs in our model of human speech production.

Although we speak of applying the prosody markup to the phonological representation, we should be quite clear that this means, for us, setting up the wrapper hierarchy which contains the utterance's underlying phonology. Prosody is logically prior to phonology–not, as is often the case in prosodic modelling, the other way round. For us an utterance's phonology is fitted to a pre-existing prosody. Notice that the flat phonological representation is fitted to the prosodic hierarchy in this method. This is a consequence of applying the prosody markup in a single processing event. We can imagine, though, that there are subsequent modifications of the basic phonology which result from an inclusion of expressive content, which in turn wraps the phonology.

Part VI

Strengthening the High-Level Model

21

Speech

21.1 Introductory Note

This section looks at models of underlying factors contributing to the structure of the speech waveform. Two major factors we discuss are biological and cognitive processes that contribute to the production of language–a medium of *expression*. In spoken language, the speech waveform is the medium of *transmission* of this expression. Speakers choose words and syntax, informed by feelings and context, modulated by what is called their *physiological setting*–their physiological state at the time. A process of cognitive intervention is able to modify the setting to produce speech in which listeners can detect expressive/ emotive content. Thus the waveform is the acoustic form of an encoding of language structure, physiological setting, and cognitive activity which combine to produce human speech with expressive/emotive speech. We dwell on this because it is a major part of what we identify as *naturalness* in speech.

Expressive/emotive content is embedded in the speech waveform. Listeners can detect its acoustic representation and speakers know how to produce it. However, to date, inspecting the waveform for acoustic changes has not been productive in identifying the expressive/emotive detail of this content. We suggest that looking at its underlying sources might help us construct a mapping from them to the acoustic events in the waveform. The goal is to make a contribution to building a high-level synthesis system which can underlie a simulation of the human waveform which itself has encoded expressive/emotive information.

So, the speech waveform is the result of a chain of cognitive and physical events based on the speaker's intention to communicate, accessing linguistic knowledge to encode the message and calling on knowledge of the ability to speak it. Speech detected by the listener is a complex mix involving a basic message (the actual information the speaker wishes to convey) and expressive/emotive content (additional information about the speaker's attitude, beliefs and emotional state at the time of speaking). Expressive/emotive information may be conveyed by choice of word and syntactic style, but much is encoded on the linguistic sub-component labelled *prosodics* (also called *suprasegmentals*). This additional information can be *intended* to evoke a particular attitude, belief or emotional state in the listener, or it can be *unintended* but nevertheless still apparent to the listener.

In addition to examining the waveform itself, we will look at the planning and production systems which underlie the detail of the waveform. The processes occurring before the final waveform is produced are described by *phonology* and *phonetics*. Production of

---

Developments in Speech Synthesis   Mark Tatham and Katherine Morton
© 2005 John Wiley & Sons, Ltd.   ISBN: 0-470-85538-X

speech–largely a physical process–is described by phonetics, and planning of speaking–a cognitive process–by phonology.

21.2 Speech Production

We observe individuals communicating with each other using speech. We notice that what they say contains both a plain message and expressive content. We can describe the production of a plain message by observing changes in the vocal tract as it is reconfigured during speaking–for example, tongue movement, velum raising and lowering, lip closure. Models of speech production are well described (Clark and Yallop 1995; Borden *et al.* 1994; Ladefoged 1996, 2001), but expressive/emotive content is not so easily characterised. In this book, the model presented relies on three basic concepts:

1 Phonology characterises underlying potential sound shapes and patterns. These are the segments making up the phonological syllable. Within phonology, prosodics characterises some overarching features of those segment sequences labelled *syllables* and *morphemes*–for example, intonation contour, sentence stress and focus. Phonology describes the intention and its result, the *plan*, to produce speech in a particular way. There are two categories of description: phonemic units and prosodic units.
2 Phonetics describes the physical characteristics of the stream of sound produced by realising or rendering the plan. It describes those changes in physical features that correlate with segments, syllables, morphemes and prosodic effects. All phonetic descriptions in this book are made assuming a correlation with some underlying phonological units. Not all researchers take this approach; some describe acoustic phonetic features with no special association with underlying linguistic units. There are two categories of description–segmental, including subcomponents such as a burst of aperiodic noise on release of stop consonants; and suprasegmental, longer term events such as fundamental frequency change during an utterance.
3 We assume that *all* speech consists of both plain message and expressive content–what has been called 'neutral speech' is a special case of expressive speech conveying very little of what we normally think of as expression (Morton 1992; Tatham and Morton 2004). The two types of event are produced simultaneously by the speaker; however, the categorisation into two types is made for convenience in model building and does not imply two separate sets of procedures. The plain message and associated expressive/ emotive events are correlated with the appropriate phonetic events and the processes that gave rise to them in order to describe the waveform.

21.3 Relevance to Acoustics

The acoustics research task is to determine which phonetic events identified in the speech waveform can be correlated with elements and processes associated with the phonological plan to produce a plain message, then determine which phonetic events can be correlated with expression/emotion. We assume it is possible to describe quantifiable features associated with the production of expressive content by observing changes in vocal cord vibration, rate of production of sounds, air pressure changes and so on (Lubker and Parris

1970; Hess 1983) and that it is possible to correlate these with phonological constructs. We assume expression/emotive effects are relatively longer time events compared with the short time events noted in production of the plain message. We also recognise it is not yet possible to determine very many physical and phonological features that can be correlated with expressive/emotive content in speech, but assume it will be possible in the near future. We assume that encoded in the waveform is *all* linguistic- and cognitive-based emotion information *necessary* for effective perception.

21.4 Summary

The functional distinction between phonology and phonetics is

- phonology is about *plans* to produce utterances
- phonetics is about the realisation or *rendering* of the *plans*.

In the description of speech production, the output of phonology is the input to phonetics (Wang and Fillmore 1961; Ladefoged 1965; Tatham 1971). In the model we suggest, 'lower phonetic units' are not derived–in the simple meaning of this term–from 'higher phonological units'. Rather two different domains are being *associated*. Phonology is logically prior to phonetics; speakers need a *plan* before speaking, just as they need to have an idea about something to say before accessing the linguistic knowledge and ability to put the idea into words. It is important to keep clear the difference between the abstract formal phonological terms that may be correlated with physical measurements, and to bear in mind that the domains of phonology and phonetics are regarded in the model as separate. Division into plain message and expressive/emotive content is for convenience in dealing with the phenomenon of this content in speech, and does not imply two separate sequential human processes or two parallel processes that merge at some point.

21.5 Information for Synthesis: Limitations

The available information needed to build a synthesis system to produce natural voice output–specifically, with expressive/emotive content–is sparse and scattered over the subject areas that describe many different aspects and multiple model levels dealing with expression and emotion: biology, psychology, linguistics, pragmatics, prosodics, phonetics, aerodynamics, acoustics. Extracting what is needed for synthesis software from all these fields to produce acceptable output has not proved feasible to date.

Development of appropriate applications models for synthesis has been hampered to some extent because the description of cognitive, biological and linguistic events from which expression/emotion are derived is formulated at different levels within the specialist subject. Describing a phenomenon at different levels entails different points of view which require at each level a different set of terms, different constructs, resulting in interpretations arising from these differing points of view and their basic assumptions.

One technique which might facilitate the mapping or rules, which could associate one level with another and thus enable the output of one level to constitute the input to another level, involves the parameterisation of the components (the sets of rules and objects at one level) and further parameterisation of objects within the components; mapping is set up between parameters rather than between composite objects. A simple example might be the interface between phonology and phonetics in linguistics, where the output of the phonology as a string of symbols *representing* the sound pattern *is* the input to the phonetics. However, even if the symbolic phonological representation is then turned into real world events–biological speech production resulting in the speech waveform–the output and input of different levels can be correlated *via* the component parametric characterisations. That is, the mapping is between parameters or features rather than between whole objects: this is an appropriate technique in current linguistics when formally relating phonology and phonetics.

Frijda (2000, p. 62) points out:

> Descriptions at higher levels cannot always be reduced to those at lower levels without loss. Likewise, higher-level categorizations cannot always be built up from the lower-level phenomena. Higher-level categorizations often, or perhaps usually, include more phenomena (for example, the nature of the emotional object or of a particular environment), as well as more interactions between the lower level-phenomena or feedback from them.

Speech researchers have the additional problem of relating constructs derived from different disciplines. One of the most difficult areas to model has been the one underlying much of the naturalness in speech: *emotion*, from which emotive content can be said to be derived. Modelling emotion is seen as a multilevel construct (Dalgleish and Power 1999; LeDoux 1996; Scherer 1993), in which the focus is on the many different ways emotion as a phenomenon can be defined and described. For speech research, as yet, models of emotion seem not to be clearly enough specified for forming the basis of a coherent model for use in synthesis. Nevertheless, if the nature of emotive speech is to be simulated to produce natural sounding speech, we have to find a way to associate biologically based (Rolls 1999), cognitively based (Lazarus 2001), and mixed models from these different domains (Clore and Ortony 2000).

22

Basic Concepts

The following is a brief description of speech production, the action that produces the speech wave, the basis of what speech researchers want to simulate.

22.1 How does Speaking Occur?

Speech is the result of aerodynamic effects resulting from movement of and within the *vocal tract*. Within the vocal tract *articulators* are involved–movable structures which enable constrictions to more or less impede airflow, or which direct its pathway through the vocal tract. We could list among the principal components or articulators: lungs, pharynx, velum, tongue, teeth, lips. In simple terms the vocal tract comprises the oral and nasal cavities. Movement of the vocal tract and its components is complex, and there are several underlying mechanisms:

- neuromuscular mechanisms which control automatic or reflex movement, as in breathing, throat clearing, swallowing etc. (a physical source and incidental to speech perhaps)
- local stabilising mechanisms which are usually reflex, such as intra- and inter-muscular feedback systems (a physical source and essential for speech)
- non-automatic *intention* to move the mechanism, as in speaking, singing etc. (a cognitive source and central to speech)
- mechanical and aerodynamic movement externally sourced, as in the passive movement of one articulator by another (a physical source important in defining the acoustics of speech).

The intention to move the vocal tract can be initiated by the desire or intention to speak, requiring the vocal tract to be configured dynamically in a particular way to produce the sounds as needed by the particular language. All acts of speaking require access to a language capability which is described by linguistics/phonetics. The particular language knowledge base for each language is shared by the speakers of that language (as is the cultural knowledge shared by the members of the culture).

Speaking also requires access to knowledge of how to programme the motor control system in order to move the articulators to produce the appropriate speech waveform. This knowledge must include

- the units of motor control–muscles with their means of neural innervation
- higher level grouping units–coordinative structures with their means of intra- and inter-communication

Developments in Speech Synthesis   Mark Tatham and Katherine Morton
© 2005 John Wiley & Sons, Ltd.   ISBN: 0-470-85538-X

- how to deal with external constraints, such as coarticulation
- how to evaluate various forms of somatic feedback and modify the innervating signal accordingly.

The speaker must also conform to the conventions of pronunciation (the phonology and cognitive aspects of the phonetics of the utterance), and to feedback from the listener as to how much the listener appears to understand what is being said.

Variability in the utterances produced by the same speaker on different occasions can be due to state of health, age and breathing rate. Among different speakers variability can be introduced mainly by the physical structure of the vocal tract, state of health, age, respiratory system etc. Completely natural synthesised speech would have to include a perceptually relevant minimum of such variability.

> We think that an additional source of variability is due to changes in the basic biological physical stance of the speaker which will be reflected to some extent in the physical shape of the vocal tract, as these changes may also be reflected in changes in arm movements, facial movements and posture. We suggest a correlation between biological changes resulting in emotion, and emotional effects appearing in the speech waveform (see also Chapter 23).

In addition, *cognitive intervention* (Ortony *et al.* 1988) can suppress or change the character of emotional content in the speech waveform as the speaker can suppress or change the expression of emotion through body language, or facial changes.

Speaking requires the use of language. We must therefore add a *linguistic element* to the coding process, including

- the fundamental concept that thought can be communicated by speaking
- a simple model incorporating the concept of basic biological stance and how general physiological state has an effect on cognitive aspects of language
- access to a lexicon of words etc. in the language for the purpose of communicating an underlying concept
- access to a phonology and cognitive phonetic components for enumerating and deriving permitted sound shapes (abstract sounds) appropriate for encoding lexical and syntactic strings.

We discuss the detail of some of these schematic concepts elsewhere; for the moment the list above is a skeletal outline of some of the main objects and processes to be included for the linguistic contribution. Speech researchers need to bear in mind that the act of speech encodes all linguistic information. This information is described within linguistics by several *components*, core among which are

- semantics, pragmatics, syntax, phonology, phonetics.

When the speech waveform is detected, it is probably the case that almost all the encoded linguistic information is interpreted. But it is important to remember that the listener adds his or her own emotional physiological state to the interpretation of the total set of information. We should also at least bear in mind cultural, social and familial elements which can be of major importance in enabling interpretation of the signal. This is important even in speech production and synthesis because it is felt that speakers actually take an assessment of the listener's environment into consideration, as when, for example, a speaker increases the precision of articulation and decreases rate of delivery in step with an increase in ambient noise.

From the description, we can see there appear to be three basic disciplines that study phenomena which underlie speech and provide useful points of view to help in characterising emotive content in spoken language.

1 *Linguistics*–a characterisation of the linguistic knowledge base, being phonological and phonetic information but also other cognitive sources such as syntactic and semantic linguistic constituents. A pragmatic sub-component, which is cognitively sourced and which may affect the choice of words and prosodic patterns in the linguistics, also needs description.
2 *Biology*–and its modelling relevant to speech. There are emotive effects which arise because of cognitive intervention in a basic biological reaction to changes in the environment. These cognitive effects can be described as enhancement, suppression, or changes the speaker judges suitable in the context.
3 *Psychology*–and its cognitive intervention potential associated with biological reactions, during an instantiation. Biologically sourced emotive content is described along with other physical properties of the vocal tract.

The linguistic information, which contains cognitive and expressive elements through the choice of words and phonological information, and some output from the phonetics processing dynamic plane, feeds into

- the motor control system . . . which then
- shapes the vocal tract which is in a biologically determined configuration . . . which then
- results in an acoustic wave.

Thus all required linguistic information is ultimately encoded on the prosodic phonetic 'carrier', incorporating segmental timing changes, delimited accent groups, intonational phrases and so on. In this oversimplified account we can see that linguistically determined properties, cognitively determined voluntary expressive properties, and biologically determined involuntary expressive properties unify in the soundwave. The convergence is in many respects a *managed process* (Morton 1992), and the final result is by definition the appropriate tailor-made trigger for perceptual processing, thanks to the principle of production for perception (Tatham 1986a; Lindblom 1990).

22.2 Underlying Basic Disciplines: Contributions from Linguistics

22.2.1 Linguistic Information and Speech

It has been generally agreed by researchers working in the 'transformational generative grammar' framework (Chomsky 1965; Jackendoff 2002) that there are two aspects of spoken language to take into account in characterising speech:

1 **what** the speakers *know* they are able to do
2 **what** the speakers *choose* to do.

There are several variants of transformational grammar, but most continue to subscribe to basic principles like this one. Note that we do not address the question of *why* speakers wish to do what they do. This depends on motivation and personality, among other things, and also interpretation of the context within which they speak; this area falls within the research of psychologists and experts in cultural context etc. (Harré and Parrott 1996).

The first–what speakers know they are able to do in general–is most appropriately characterised by a *static* model. But the second–what unique or local choice is made on any one occasion–needs describing as a *dynamic* model. The contents of the static model consist of characterisations of speakers' knowledge of their language. Choosing to draw on this knowledge to produce a unique and appropriate utterance is a dynamic process and depends on context, on the speaker's physical ability, and on the speaker's awareness of the most suitable speech for the occasion, especially important in responding to ongoing changes in a conversation. That is, *knowledge* is generally characterised in static models, *action* in dynamic models. (See Part VII for a fuller description of static and dynamic models.)

The discipline that describes possible language structures underlying the production of spoken language is *linguistics*, and core linguistics is generally concerned with a static model characterising knowledge of the language as a set of explicit statements. Linguistics is generally sub-categorised into components–semantics, pragmatics, syntax, phonology and phonetics–and descriptions in these sub-categories aim to be formal and explicit; this is essential for adequate computational modelling of language. The phonetics component falls slightly short of fully explicit descriptive adequacy as, we shall see later.

Models of language within or derived from the traditional transformational grammar framework deal with the structure of language (knowledge) but not with how its structure produces real-world sentences (instantiation). In most contemporary models the characterisations of choices being made on any one occasion are in the domain of psychology, psycholinguistics and phonetics (but not semantics, syntax or phonology).

For speech research, a speech waveform can be labelled using these linguistic descriptors. One problem is that the assignment of phonetic labels in particular depends to some extent on interpretation, the exercise of intuition, and the knowledge and experience of the person assigning the labels. Because the assignment is not totally rule-governed, there is often a difference of opinion which results in a certain amount of variability in the description of even a simple speech waveform. Assignment of other labels, such as grammatical categories and syntax, is usually applied in a more standard way.

Labelling expressive or emotive information is not as formal and explicit as linguistic labelling. It has not been possible to determine where emotive content occurs in the waveform, nor possible to label emotive content as well as language content can be labelled. Thus a fully specified applications model consisting of emotion terms and labels on the waveform, needed for building a synthesis system, has not been made.

22.2.2 Specialist Use of the Terms 'Phonology' and 'Phonetics'

For the purpose of establishing a firm basis for speech synthesis, the most important areas of description within linguistics are phonology and phonetics.

Phonology characterises all possible sound patterns in the language, and in doing so it describes

- phonological plans for the sequencing of speech sounds and segments making up the phonological syllable and ultimately entire utterances
- overarching prosodics like intonation contours and rhythm, and word and sentence stress.

The traditional domain of phonology–the sentence–can be enlarged to the paragraph, or include an entire text and include sub-domains down to segment length. Input to the phonology is derived from the syntactic component in traditional approaches to transformational grammar, the syntax being responsible for the appropriate ordering of words or morphemes within the sentence domain. As part of the overall static approach to linguistics, phonology has no time-dependent element. Speech is the result of the realisation or rendering of phonological plans.

Phonetics describes

- speech production processes
- the nature of the acoustic output
- some areas of perception.

Phonetics characterises the vocal tract configuration at any one instance of speaking and the sequencing in time of these configurations. In doing so it accounts for ongoing changes in physical features that are correlated with both phonological segments and the prosodics the segments occur in. Cognitive phonetics (Tatham 1986b; Tatham and Morton 2003) characterises the type of knowledge required to form the correct configuration to render the phonological plan, and, importantly, models the supervisory processes in speech production.

22.2.3 Rendering the Plan

The term *phonetic prosodics* refers to physically observable phenomena that overarch segments, such as fundamental frequency change and durational changes in segments like individual sounds and syllables, that can be correlated with the prosodic framework for plans. Many of these phonetic descriptions are of short-term events, such as individual segments. Longer term events that continue beyond the short-term articulation of segments are also described. Examples of these are

- vocal cord vibration
- rate of production of the utterance
- varying airflow and air pressure changes
- some jaw and muscle movements.

The absence of clock time in phonology is due to its cognitive nature. Unlike phonology, phonetics must be time-dependent because of its physical nature.

The difference between short-term events and longer term events is important since language characterisations, explicit or implicit, are in the form of a hierarchal set of descriptions, implying that longer term events dominate shorter term events; that is, the latter occur *within* the former. In the case of speech, all phenomena can be described as a series of short-term events, within a longer time *frame*. A sound of short duration must traditionally, just like any lengthy sound or collection of sounds, be linguistically meaningful. A single short burst of sound, such as *ugh!* produced by an individual, is not classified as speaking. And some measurable short bursts of sound can be produced only in conjunction with other sounds in a longer stretch; stops or plosives form the class of segments usually cited here. Stops require a vowel sound or vowel-like sound (such as [l] or [r]) either before or after the air flow is stopped; sometimes a brief moment of a-periodic noise can follow a stop, as in Cockney English slow release, for example [t] in a word like *tea*. Expressive/emotive content in speech involves longer sequences of sounds because of its prosodic phonetic level of rendering.

22.2.4 Types of Model Underlying Speech Synthesis

The type of model is central to understanding the linguistic underpinning of utterances, carrying over eventually to their simulation in synthetic speech. Two types of model–*dynamic* and *static*–are basic to the approach to synthesis, we suggest (Tatham and Morton 2003). Dynamic and static modelling are relevant to both phonology and phonetics.

The Static Model

Static models are designed to account explicitly for *all* possibilities in the language. They are expressed as a generalised data structure. The static approach is entirely compliant with the core transformational grammar approach to syntax which accounts for all possible sentences in the language–never, as a component, for just one sentence. As with syntax, exemplar derivations are possible (Tatham and Morton 2003), but they are always just that–examples from the entire set of characterisations. Static phonology, for example, enumerates all possible utterance plans, and static phonetics accounts for all possible renderings of phonological plans. Thus we say that the *potential* for all utterances is described in the static phonological and phonetic models. This is in sharp contrast with the description of a real-world speech waveform, a *particular* utterance which is an instantiation 'selected' in a motivated way from all possible sound shapes and their potential phonetic renderings.

The Dynamic Model

Dynamic models include an account of selection processes whereby individual plans of utterances are derived from the set of generalised data structures on the static *plane* of phonology. Note the use of the term 'plane'. We envisage two planes, static and dynamic, housing their respective sets of elements and processes. Phonetic features and processes appropriate to the *particular* utterance, described in its plan, are selected from the phonetic inventory on the static plane (where all possible renderings are characterised). Rendering gives instructions to the motor control system to shape the vocal tract so that the appropriate speech soundwave is produced. The rendering is constrained by the physical emotive state of the speaker and by pragmatic considerations, regulated by the 'cognitive phonetic agent' (see Chapter 27).

Dynamic models characterise an individual plan, and how the utterance described in that plan will be spoken. The characterisation takes the form of an appropriate data structure and procedures retrieved from the static phonological and phonetic planes. All plans for all utterances are characterised on the phonological static plane. The static phonetic plane contains information about the speaker's ability to render all plans. A particular utterance plan is assembled on the static phonological plane and rendered on the dynamic phonetic plane. There is more detail about the model in Part VII, but the following diagram illustrates the distinction between static and dynamic planes, and between phonological and phonetic processes.

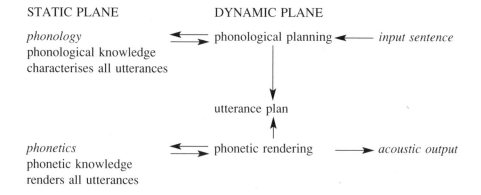

STATIC PLANE   DYNAMIC PLANE

phonology   phonological planning ◄——— *input sentence*
phonological knowledge
characterises all utterances
                                                │
                                                ▼
                                      utterance plan
                                                ▲
                                                │
phonetics   phonetic rendering   ———► *acoustic output*
phonetic knowledge
renders all utterances

23

Underlying Basic Disciplines: Expression Studies

23.1 Biology and Cognitive Psychology

We think that some understanding of the immediate precursors of the soundwave might help researchers identify expressive and emotive content of speech and understand the prosodics used to convey them. Even a quick search of the relevant literature in biology, cognition, psychology, linguistics and cognitive science produces a vast amount of research on expression/emotion. In this book, our perspective is based on a narrow selection of this work. A more detailed discussion based on the research literature, again selective, but directed toward a perspective on cognitive and biological models for expression in speech research in general, can be found in Tatham and Morton (2004).

Most of the literature focusses on emotion, and so we deal with this aspect of expression. Emotion is particularly interesting since it can be thought of, as we shall see, as being dual-sourced, with origins in both biology and cognition. Since we need to label emotion before correlating it with labelled acoustic features, let us look at what is generally accepted by emotion researchers. Emotion is usually considered to be a biopsychological *event* (LeDoux 1996; Damasio 1994; Panksepp 1998; Plutchik 1994; Scherer 1993; among others). However, researchers in the field of emotion studies are not agreed on what actually constitutes emotion.

Biological evidence provides a good and sufficient explanation for the source of emotion for some researchers (LeDoux 1996; Rolls 1999). Others concentrate on cognitive aspects (Averill 1994; Frijda 1993; Lazarus 2001), and yet others regard emotion as sourced both cognitively *and* biologically (Adolphs and Damasio 2000; Borod 1993; Davidson 1993; Johnstone *et al.* 2001). Within this last group, there are those who regard emotion as primarily cognitive with some biological base (Frijda 2000), and others who treat emotion as biologically sourced with the potential for cognitive intervention (Panksepp 2000). Whatever the point of view, there seems to be general agreement that emotion can be regarded as an *event* experienced by an individual who may be aware or unaware of the event at the time it occurs (Ekman and Davidson 1994; Frijda 2000; Zajonc 1980).

Developments in Speech Synthesis   Mark Tatham and Katherine Morton
© 2005 John Wiley & Sons, Ltd.   ISBN: 0-470-85538-X

23.2 Modelling Biological and Cognitive Events

Traditionally, modelling strategies for biological and cognitive phenomena are assumed to be different since they are thought of as being *qualitatively* different (Damasio 1994; Descartes 1649; LeDoux 2000; Panksepp 2000). Although researchers are aware of the reality that a human being acts as a single unit, the division into two types of phenomena has been useful. Therefore, along with many researchers in the field, we will consider the emotive effect in speech as being produced by two underlying events that may occur simultaneously yet be described according to different principles and investigated by different methods. In doing so we take into account

- a *cognitive contribution* to spoken language
- a *biological contribution* to spoken language.

One major obstacle in building these models is that access to biopsychological processing is usually possible only by indirect means; coherent models seem to rely on cautious inference from less than ideal experimental work. Some animal experiments as well as observations and descriptions from human disorders can provide useful information. To date, however, it has not been possible to construct a full model of emotive content that can be reliably associated with the speech waveform. It follows that speech researchers will be limited in the confidence they can place in making these associations.

23.3 Basic Assumptions in Our Proposed Approach

The basic assumption in our own approach is that the interaction of the biological and cognitive domains is the source for emotion. The result of this interaction is to signal the neuromuscular system to move the vocal tract into a configuration the speaker *intends* and which will realise the plan to communicate. Vocal tract configuration can be shaped by non-cognitive *unintended* events as well. For example, anger can produce overall muscle tension, including tension in the vocal tract, which will change the shape of the vocal tract configuration to an extent that the resulting speech waveform will contain information the listener can detect and recognise as the emotion *anger*. It is also possible, as another example, to shape the vocal tract intentionally by consciously relaxing muscles in such a way as to produce the perceptual effect of *calmness*. A major point we wish to make is that emotion can contribute both to *intended* and to *unintended* movements in the vocal tract. The perception of such movements by a listener depends on a threshold factor, of course.

23.4 Biological Events

It is possible perhaps to associate biological events with speech waveform acoustic events. For example, a time-governed dynamic model of the airstream mechanism can be appropriate for, say, stop consonants. Further back in the chain of production events we can, for example, detect and assess

- the results of neuromuscular activity using electromyography (MacNeilage 1963; Tatham and Morton 1969)

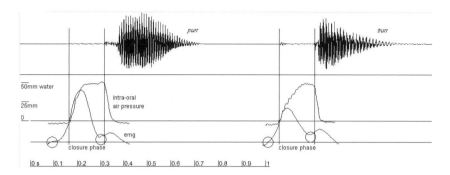

Figure 23.1   Waveform, intra-oral air pressure and rectified/smoothed electromyographic traces associated with the two isolated words *purr* and *burr*. The electromyography is from m. orbicularis oris which rings the lips and contracts to achieve lip closure for bilabial stops. For each word the period of lip closure is shown, and is clearly apparent from the waveform and air pressure traces. Note, though, that the EMG signal begins appreciably before closure is actually achieved, so that the mass of the lips can move (in this case) from a neutral position to full closure sufficiently tense to support the increasing air pressure. Note that the EMG peaks well before lip opening, indicating the start of slackening of the lips ready for them to be pulled apart at the desired moment of release. There are always several muscles contracting to achieve any one particular vocal tract configuration; their contractions are not simultaneous, but the movement of the associated articulators achieves the goal of synchronicity associated with a particular sound segment. This figure has been adapted from Tatham and Morton (1972).

- air flow and air pressure using aerometry (Lubker and Parris 1970; Tatham and Morton 1972);
- vocal cord activity by electroglottography (Hess 1983)
- voice quality by spectrography (Epstein 2002).

Such physical descriptions, which can involve *correlation*, are of changes in the neuromuscular system or the immediate effects of such changes (e.g. air flow). But because *labelling* the waveform is not rule-governed, and assignment of a label depends to some extent on inexplicit experience, the error in associating biological and physical events with labels may be quite large. Figure 23.1 is an example of a waveform marked for some of these properties. Notice the apparent lack of synchronicity between the symbolic label and associated events.

Obvious biological changes have been associated with longer stretches of acoustic changes in the waveform. For example:

1 Increased breathing rates can be associated with relative changes in fundamental frequency correlating with varying rates of vocal cord vibration.
2 Formant values (both amplitude and frequency) can be seen to vary according to increased rate of delivery. Longer or shorter vowels mean changes in onset and offset in the absolute frequencies and amplitudes of formants as well as in their relative values.
3 Stop and other consonants may be shortened or lengthened in duration when there are changed rates of delivery in general–but not in any obvious direct correlation to the syllable nuclei they envelop.

So, for example:

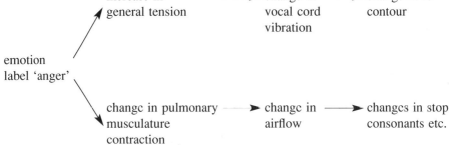

We can see by inspecting a spectrogram of speech judged to have 'angry emotive content' that there are differences compared with non-angry speech. The duration of the stop (silent phase), and/or the amplitude of the burst, the timing of the following vowel, and the formant values on vowels can be correlated as shown in Figure 23.2, spectrograms of the

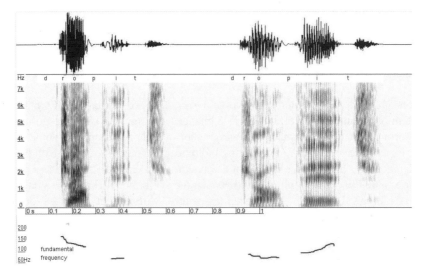

Figure 23.2 Waveform, spectrogram and fundamental frequency curve of the phrase *Drop it!*, spoken angrily (left) and calmly (right). Note the differences in rate of delivery (overall and within the utterance), amplitudes of the various segments, and the direction of fundamental frequency changes. We know something of the gross acoustic differences used to convey differences of expression, but little as yet of the more subtle details involved with emotions less extremely contrasted than anger and calmness.

phrase *Drop it!*. Establishing invariances to act as a baseline against which to assess these differences has, except in the most general way, proved unsatisfactory.

Emotion is not, of course, the only feature to be conveyed by differences in baseline prosodics. For example, in no particular emotive context, speakers regularly introduce features such as *contrastive emphasis*, as in a sentence such as *I read newspapers, not magazines*. Here, local modification of the intonation contour and perhaps also the patterning of the stressed syllables results in changes in the expected fundamental frequency contour, signalling to the listener that the speaker is making a deliberate comparison between newspapers and magazines. This is described as 'linguistic contrastive emphasis' (Jackendoff 2002).

Varying biological conditions of this type have been used to differentiate some emotion types. Relationships are complex and not always understood; but we raise this point here to illustrate that improvements to the naturalness of synthetic speech will depend on subtle modifications to the acoustic signal, which in human beings are sourced from different underlying factors–not all of which correlate in a linear fashion. We believe that improved understanding of underlying biopsychological factors will greatly assist the successful introduction of expressive/emotive content into synthetic speech, thereby moving forward quite significantly its naturalness and therefore its usefulness.

23.5 Cognitive Events

A major concern of many researchers is the possible relationship between cognition and emotion. They put forward a major functional construct called *appraisal*. Appraisal operates as an element of *cognitive intervention* after the basic biological reaction associated with emotion (Scherer 2001). In this view, cognition is thought of by some researchers as the source of the emotional experience (Oatley and Johnson Laird 1987; Ortony *et al*. 1988); this cognitive activity may be below the level of consciousness. The main function of appraisal is to identify an event and to evaluate to what extent it is significant for the individual (Ortony *et al*. 1988; Scherer 1993; Lazarus 2001).

Appraisal as a cognitive event can influence behaviour, including language response in the speaker. A language response is specific to the individual at the particular time of speaking. The response can be encoded in words, grammatical constructions etc. to provide emotive content.

An emotion response may result in an overall biological effect which will also appear in the waveform as a change in the acoustic characteristics of the signal. We think appraisal is a useful concept when characterising a cognitive source of emotion that may ultimately be encoded in the speech soundwave. Such a source may influence choice of words, and introduce other linguistic phenomena such as the contrastive emphasis alluded to.

Ortony *et al*. (1988) focus on the cognitive precursors of emotion, and present an analysis that can be seen as a computational approach. They outline a model in which cognitive antecedents play a causal role in emotion experience. The relationship can be diagrammed:

a cognitive component ──────────────────────────▶ emotion experience

has a direct effect on

The concept of appraisal is a major constituent of the cognitive component for these researchers. They refer to the point at which the cognitive component effects a change in the individual (so that he or she experiences emotion) *as the cognition–emotion interface*. The model specifies a set of emotion types in terms of the conditions under which the emotions are elicited. Thus:

certain conditions ⟶ different emotion types
can elicit

In a slightly different formalism we might say:

a. emotion ⟶ emotion *type 1* / condition *X*

is evoked as   *in the environment*

b. emotion ⟶ emotion *type 2* / condition *Y*

The environment condition specifies 'what is necessary, but not always sufficient, for the tokens associated with a particular emotion type' (Ortony *et al.* 1988, p. 172). To paraphrase Ortony *et al.* (p. 178), their model–simplified for our purposes–is as follows:

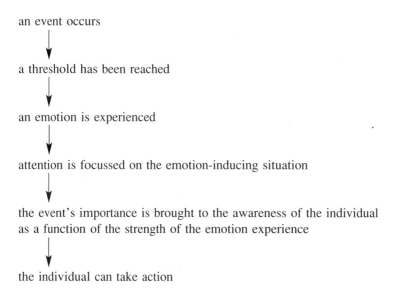

an event occurs

a threshold has been reached

an emotion is experienced

attention is focussed on the emotion-inducing situation

the event's importance is brought to the awareness of the individual
as a function of the strength of the emotion experience

the individual can take action

One of the conclusions they suggest is that the strength of the physiological reaction may be twofold:

- to focus attention on the situation
- to prepare the body to take action.

However an emotion is sourced or characterised, its function is seen by many researchers as

- to focus attention on an important event (Ortony *et al.* 1988; Niedenthal *et al.* 1999; Öhman *et al.* 2001)
- to provide information to the individual (e.g. Ortony *et al.* 1988) about the self and the environment
- to index memory to enable better responses to changes in the environment on the basis of past experience, and success or failure in dealing with a similar event (Tomkins 1979).

According to some researchers, including Clore and Ortony (2000), emotion arises as a *result* of establishing situational meaning. They suggest that

> an emotion is elicited when one's perception of a situation matches the deep structure of situational meaning that defines emotion (p. 41);

> the cognitive claim is that emotions are reactions to (or representations of) the personal meaning and significance of situations, not that emotions originate in the cerebral cortex (p. 42).

In this model, emotion can be seen as integral to our interaction with the environment. Our responses to some extent (including speech) may be directly related to the presence of an emotion reaction.

1 Communication through language will be influenced by the nature of the emotion generated.
2 Communication of emotion through language and other media will be unique to the individual.

One consequence for speech research is to model how to capture the same level of *individuality* in an artificial voice that the listener may expect from a natural voice. In the future, we suggest researchers might include fitting this into an XML or some other suitable hierarchical framework, and introduce this concept by means of an attribute tied to 'eliciting conditions'–for example, an external stimulus, or another emotion already produced. Such attributes could take on values, either numerical or verbal.

23.6 Indexing Expression in XML

Based on the work of Ortony (1988), Scherer (2001) and Frijda (2000), we might need to specify, with respect to the reaction and response of the individual, each emotion indexed with the following attributes:

1 *awareness* (yes/no)
2 *attention* focussed on an important event (degree)
3 *index memory* (efficiency rating)
4 *thresholding importance* (degree)

5 *underlying cognitions* (attributes)

6 *potential appraisal*–Is it possible? Do underlying cognitions exist–as proposed by, for example, Ortony *et al.*–and is evaluation possible? (yes/no; degree)

7 *learned behaviour* possible (yes/no)

8 individual can *experience* it (yes/no)

9 individual can *react to* it (yes/no; degree)

10 unambiguously *differentiable* from others of equal status–that is, capable of being categorised uniquely (yes/no)

11 *capable of cognitive intervention* (yes/no dependent on strength of stimuli–that is, degree)

12 *sensory experience*, and capable of sensory input (yes/no; degree)

13 *biological reaction* (yes/no; degree)

14 *behavioural response*–and suppression of response (yes/no)

15 potential for being rated '*significant to the individual*' (yes/no; degree)

16 *variability* (numerical range)

17 potential for new events *triggering prior experiences*, whole episodes, not just generalizations from several experiences (yes/no; degree)

18 potential for *calculating from novel events* (yes/no; degree)

19 can be *triggered by mental representations of an outcome* (yes/no)

20 can itself *trigger rapid action* (yes/no)

21 can itself also *enable flexibility* (yes/no)

22 can be *caused by appraisal* (appraisal model assumed) (yes/no)

23 '*reinstatement*' and '*computing a new significance for an totally new event*' (appraisal model assumed) (yes/no)

24 can function as *providing information* (appraisal model assumed) (yes/no)

25 can *form hypotheses* about others' behaviour (yes/no)

26 can *be communicated* (yes/no)

27 capable of *being elicited* (yes/no)

28 capable of *changing from one emotional state to another* because of change in environment (yes/no).

The above set of attributes, although not exhaustive, already looks somewhat daunting. In terms of speech synthesis it is clearly premature to contemplate anything like a full implementation. Rather, we have introduced this list to show just how far the area of modelling expressive content has progressed in just this sub-part: emotion. Speech synthesis researchers will need to consider how these contribute differentially to naturalness by experiment, and there is no doubt that this will need considerable research investment in the future. Work on human speech can be included in a supporting model (Tatham and Morton 2004).

23.7 Summary

If we are to characterise emotive content of speech, we need to ask these questions:

1 Can a formal model be stated that characterises types of emotion which potentially can be related to acoustic features?

2 Can labels be assigned to emotions to differentiate them as types?

And for speech research, we need a set of defining attributes and unique values which can differentiate one emotion from the others. Simulating a speech waveform with emotive content is not simple, bearing mind that the waveform acquires expressive content when

- a complex set of differentiated biological reactions occurs when a change in the environment occurs;
- cognitive processes verify the change, put it in context, and appraise the significance of the stimulus and of the reaction for the individual;
- appraisal as a process occurs before the linguistics component to enable the speaker to select the correct word or phrase, and linguistic features such as contrastive emphasis.

24

Labelling Expressive/Emotive Content

So far researchers have been unable to provide a satisfactory way of labelling expressive content in the soundwave, at the same time characterising the derivation of that content. We have to ask ourselves why this is the case. Ideally, for each identified and labelled emotion, we would have a file specifying the parameters characterising them and identifying their sources. A unique specification within each contributing domain would be defined, and then linked sequentially to show the complete derivation of expressive content cascading from cognitive contributions, through phonological, phonetic and biological factors to the final acoustic waveform. So far such a comprehensive characterisation of the derivational history of acoustic features contributing to expressive content and its perception has not been possible in any useful way. We suggest below several reasons why this is the case.

1 No unique definition of expression or emotion exists which can reliably be correlated with acoustic parameters such as fundamental frequency, overall amplitude or intensity, formant frequencies, amplitudes or bandwidths, phone or syllable duration, or rhythm–in any way traceable through the above domains.
2 The phenomenon of emotion is a reported *experience* by an individual and regarded as a single event. Characterising the experience, the extent of awareness of the experience, analysis of the context and conditions under which this experience occurs, the reliability of reporting by the individual and the accuracy and relevance of observation of the behaviour of an individual, the biological correlates–are all researched and discussed, though from many points of view.
3 Given the complexity of the event, it is clearly not possible yet for speech researchers to correlate multi-level models from other disciplines with the acoustic features of speech.

As an initial step, let us broadly outline some areas of research that seem to be relevant to understanding expressive/emotive speech.

- *Biological domain.* Research suggests that four major neural circuits can be isolated which may be associated with four types of emotion reaction (Panksepp 1998).

Developments in Speech Synthesis   Mark Tatham and Katherine Morton
© 2005 John Wiley & Sons, Ltd.   ISBN: 0-470-85538-X

- *Cognitive domain.* Appraisal models exist which outline the area of cognitive emotive investigation (Frijda 2000; Lazarus 2001; Oatley and Johnson-Laird 1987; Scherer 2001).
- *Biological and cognitive domains interrelated.* This encompasses suggestions about looking at biological events which are modelled with cognitive intervention (Davidson 1992), or looking at cognitive events which are assumed to have a biological substrate (Ortony *et al.* 1988). This does not mean that cognitive events are derived from this substrate– cognitive models are handled quite differently (Bechtel and Mundale 1999).
- *The lexicon.* The emotion experience is reported with words. Some experimental stimuli are formulated in terms of emotion words, and the results of the experiments are described using emotion words. Problems have been noted in using the same code to describe the recognition and identification of a phenomenon and to describe the work of researchers' experimental set-ups and their description of results (Borod *et al.* 2000; Harré and Parrott 1996; Wehrle and Scherer 2001).

These domains are large and therefore require a fairly narrow focus within each domain for productive and meaningful research. Inevitably, the mind-set of researchers may influence the type of research conducted and recognised. As Frijda (2000, p. 72) points out:

Perennial problems are often not resolved because they reflect a particular world view or limits in capacities for conception.

We echo Frijda and others working in this highly complex area to which so many disciplines, each with their own meta-theories, contribute. It is indeed not infrequently the case that a suggestion for a paradigm change is not accepted immediately by the general research community, and this may slow the dissemination of important new ideas. As has been pointed out by researchers in each of the fields we mention, this is particularly irksome in interdisciplinary work, where model building may be based on sound principles from one discipline, but calling for flexible interpretation in another. Returning to the perspective of this book–the improvement of current speech synthesis–an example of improvement might be to recognise that the goal of a synthesis system is to trigger an appropriate response in listeners, rather than to replicate speech waveforms. This is the approach taken here because we believe that, just as a human speaker must be considering the perception of utterances as he or she speaks, so speech synthesis systems must incorporate a model of the listener if they are to come even close to their output being reported as natural.

24.1 Data Collection

The most basic models of computer simulation of expressive/emotive content in speech are based on a simulation of the output of the human speech production process, and not on the underlying processes that produce the speech. There are some models of the articulatory processes that produce the speech output, and some investigations of the human cognitive production system. There are many excellent reports of experiments on physical aspects of speech production. But the most useful for the purpose of synthesis are those that are computationally oriented, since they explicitly lead into processes which we might incorporate in synthesis algorithms. We discuss just one of these here as an illustration of how data might be gathered for the purpose of computer-oriented modelling in speech.

Wehrle and Scherer (2001) have built a computer-based system which elicits a response from subjects in behavioural tests when presented with a fixed set of words. Subjects are asked to report which word can most closely be associated with what the subject perceives as his internal state. The objective is to test the plausibility of appraisal theories of emotion. The principle behind the design is that, from data gathering of this type, reporting of experience of emotion and assignment of a label to that experience, a mapping can be made from stimulus input to word assignment output.

Their goal was to formalise an approach intended to enable a computer to reason about emotion simulations of natural human processes. It has to be this way round because it is generally accepted that machines could not currently be said to experience emotion. A successful machine implementation would also validate the theory on which the implementation is based, and test some of the basic concepts behind appraisal theories of emotion. They clearly separate the model from the production of a machine with affective features:

> An important distinction has to be made between theory modelling on the one hand and artificial emotions on the other. (p. 350)

The varying bases for responses (named *causal input factors*) must be specified, along with structures that will allow this change, and the relationship between the structures. They ask whether one particular structure requires or presupposes another, with questions like *Are any structures optional?* and *Are any restrictive?*

In this book, we are not proposing a way of building a machine whose purpose is to reason about emotions–that is, constructing artificial emotions. But outlined in Wehrle and Scherer we see a potentially useful way of obtaining data that may characterise processes associated with input and output from reports by individuals. We repeat that it is precisely this kind of research which is needed to push forward our own particular application in the form of increased natural synthesis.

At first sight an experiment presenting a set of words as stimulus items predicted to elicit a verbal response would seem simple in outline:

stimulus word ——➤ emotion ——➤ response word

But in conducting a practical experiment, there may be unexpected and unexamined variability in the assumed processing. Wehrle and Scherer (2001, p. 354) state the problem as:

> One needs to specify the effects of causal input factors on intervening process components (often hypothetical constructs) as well as the structural interdependencies of the internal organismic mechanisms involved.

24.2 Sources of Variability

Just from looking at the approach outlined here, cited because it is computational, rigorous and process-oriented, we elaborate on the problem for the researcher who is limited by the variability inherent in the human subject, and has to make a number of complex and interrelated assumptions:

1 The stimulus items and the behavioural responses can be associated.
2 The subject understands the meaning of the word in the same way that the researcher intends it to be understood–that is, shared connotation and context. This assumes a shared meaning describable as a prototype (Lakoff 1987; Niedenthal *et al.* 2004).
3 Differentiable emotions occur.
4 The stimulus words refer to differentiable emotions.
5 The subject does not suppress the emotion (because of a special meaning) or suppress some other emotion that might arise within the individual. See Clore and Ortony (2000) on emotions giving rise to *other* emotions.
6 The subject does not edit the response, and provide the researcher with what the subject thinks the experimenter wants or will accept.
7 The subject reports accurately–that is, within the experimental paradigm.
8 If a forced choice is presented for response, the words chosen are words or phrases the subject understands; and the subject accepts the correspondence between these words and the emotion elicited.
9 The subject is not confused, frightened, bored, distracted from the experimental situation (which can happen in some cases during experiments involving organism responses).

Thus the emotion researcher may be forced to rely on subjects who are sensitive to the object of the experiment: emotion. Subjects are aware of being asked to report some aspect of emotion response; they may be asked to avoid editing their responses, to avoid confusing the meanings of words used both as stimuli and, in some experiments, as response choices, but may be unable to do so.

Although emotion models are incomplete, what *has* been modelled can be associated to some extent with acoustic labelling of events in the waveform. For example, we can measure acoustic events, even if there is some disagreement about the predictable assignment of emotion labels (Murray and Arnott 1993). These events include the often cited changes in fundamental frequency contour, formant structure, amplitudes and duration.

A foundation for stability in fathoming the acoustics of expression is the fact that *many* acoustic events can be assigned an acoustic phonetic label. Within the linguistic framework, phonetic labels can be related to phonological labels, to syllables, and finally to words. At the word level, a semantic meaning can be associated, and the place of a word in the grammar can be labelled (category and syntax). Thus, a linguistic specification can be built up and associated–although not perfectly–with the acoustic *terms*. Techniques for gathering data can be worked out, but the main problems lie in constructing hypotheses, and in evaluating, generalising and modelling the events subsequently observed. The weak point is the initial association of an emotion *label* with the acoustic *event*. 'Finding' an acoustic event has not proved successful, and the strategies for labelling the acoustic signal are not robust. It is for this reason, among others, that we suggest looking at underlying processes in producing expression/emotion with a view to enabling a more hypothesis-driven approach to determining the nature and perceptual significance of acoustic events in the speech waveform.

24.3 Summary

We see emotion–understood as an exponent of expression–as the result of psychobiological processes, and we follow other researchers in assuming it to be reasonable to partition these processes into two domains, biological and cognitive, within one individual. Thus emotion is the result of biological activity, cognitive activity, and the effects one domain can have on the other. The individual can show the presence of an emotion event by a change in behaviour: reporting it, expressing it for example within an art form, or through body movement, including speaking. The individual may be aware or unaware of the expression, of the reporting, and of body movement. The task for the researcher modelling emotive expression, as we see, is considerable.

25

The Proposed Model

The basic assumptions we make are that expressive/emotive content

1 is produced by a biocognitive process
2 is the result of differentiation processes that have occurred within the individual
3 appears in the speech waveform as a result of these processes
4 can be identified and conveyed by assigning language labels
5 has as its purpose the triggering of listener percepts
6 can be described as an XML data structure.

It should be remembered that we are using emotion as an example of expressive speech because this is where so much of the work has been done, but our proposals about modelling extend to all forms of expressive content. A fuller description of the cognitive and biological correlates that might be associated with processes that produce emotive content of speech is summarised in Tatham and Morton (2004).

25.1 Organisation of the Model

The following observations have been made about the speaker's role in spoken language. He or she knows

- *what* he/she wants to do (speak)
- what he/she *can* do (has the ability to do so)
- *how* to do it (recruits the motor control system and relevant controlling or managing cognitive structures)
- *how* to respond to ongoing changes in response to the listener's responses (he/she responds to feedback from the perceiver).

In accounting for production we distinguish between knowledge and instantiation:

1 *Knowledge*–including an ideal example of the implementation of this knowledge; that is, the competence and derived exemplar model based core linguistic theory.
2 *Instantiation*–implementation of the knowledge of how to produce speech and the real world production of a speech event.

Developments in Speech Synthesis   Mark Tatham and Katherine Morton
© 2005 John Wiley & Sons, Ltd.   ISBN: 0-470-85538-X

25.2 The two Stages of the Model

We suggest building a two-stage model to include characterisations of the following:

1 *knowledge*–output: *plans* of what can be done
2 *plan rendering*–output: *speech events.*

The speech output matches the intention of the speaker to use his or her knowledge of the language and communicate in a particular way. It is possible to model human speech by

- a mainly declarative model for statements about knowledge (includes data structures for developing plans and rendering them)
- a mainly procedural model for instantiation (includes procedures for developing plans and rendering them).

The intention is that data structures be characterised within a set of XML documents. Notice that the word *document* means an integrated set of descriptions, not a paper or screen document (a concept within the very narrow use of XML for text and image layout–a use irrelevant to our purpose).

25.3 Conditions and Restrictions on XML

What does XML allow us to do? What are the restrictions?

1 XML usually defines declarative statements. Such statements specify, in our usage of XML, *what* is needed to be known by the speaker, and forms the basis for constructing a knowledge base.
2 The declarative statement is used in characterising the speaker's knowledge. These characterisations are modified and often amplified by sets of relevant attributes. Knowledge is not just a set of facts (that is a plain database) but rather includes attribute statements about conditions, modifications and consequences.
3 XML is restricted to declarations, features, stating potential actions, naming classes of object; it is not primarily designed to characterise processes, and is not able to do so without enhancements not in the spirit of the W3C specification. (In this sense XML is similar to PROLOG (Clocksin and Mellish 1994) which was good at formulating declarations and poor at formulating procedures.) We need to specify

   - variable events as input, which can affect processing procedures (together with their resulting effects)
   - how the mechanisms involved work in a single instance
   - the function of the structures in the system which interrelates these mechanisms.

4 We use XML as the basic formalism for making declarative statements about the knowledge needed for speaking, and a pseudo-code to suggest ways of describing an instantiation resulting from drawing on this knowledge to differentiate expressive/emotive content. We are characterising abstract processes, accessing knowledge bases in the spirit of linguistic competence, *not* the actual performance of the individual.

As we have pointed out, expression takes many forms in the individual–writing, dancing, playing, painting, sculpting, model building and theory construction, speaking, and so on. It seems reasonable to assume that the desire to be expressive is evoked prior (logically and temporally) to actually doing something to express. We assume emotive content of speech to be a type of expression, and indeed that emotive content can be seen in other types of expression–writing, graphic art etc.

For example, a sketch of the general context provided by expression written in XML is

```
<expression>
        <writing/>
        <dancing/>
        <speaking>
              <linguistic_module>
                      <prosodic_module/>
              </linguistic_module>
        </speaking>
</expression>
```

The linguistic module specifies the knowledge needed to speak the language, results of choosing words, syntax that may contain emotive content such as *I always feel happier when the sun shines*, and so on. This module *contains* the prosodic_module which interprets cognitively sourced expression/emotion; for example, emphasis on the word *always* in *I'm always nervous when crossing this road*. The linguistics module contains also the phonetic inventory and the cognitive phonetics component which manages the rendering.

25.4 Summary

In building a computationally adequate model for the purpose of synthetic speech, we want to deal with multiple descriptions on different levels, and characterise them within the framework of a formal data structure. Statements about underlying biological, cognitive and linguistic systems that produce expressive/emotive speech need to conform to the rules of use of XML. Constructs needed by psychologists or biologists, and not necessarily developed with our purposes in mind, may well be contorted by the requirements of the XML system. Given these possible restrictions, we outline an approach which allows us to declare the terms used to describe speech, to state what conditions allow choice, ranges of values required (not the values themselves), and optional objects etc. within the XML structure (see Chapter 30). Following an XML set of declarations, a procedural coding is necessary for producing single instances (instantiations). We use a *pseudo-code*, a set of 'if . . . then' statements, to illustrate points made in this book.

On a practical note, it may be that synthesis designers will not need to distinguish very many different types of emotive content. There is also a level of granularity of representation beyond which for the purposes of synthesis it would be unnecessary to completely specify the underlying categories and processes. For example, biochemical information would probably be of no importance; but for an applications model in a different field, say, for mood disorders such as depression, a biochemical specification could be basic in correlating chemical changes with reported emotional tone and observed behaviour changes.

26

Types of Model

We assume that emotions can be differentiated–human speaker/listeners do this reasonably well, although mistakes do occur. In modelling differentiated emotions, we can build models in two ways.

1 *Category*–consists of statements about an object or event. When instantiated, features would be on/off for any given moment in time. A graph can be plotted of how the feature changes with time, and time aligned with other features as the event unfolds.
2 *Process*–consists of statements about what is underlying the object or event, not the event itself. This specifies the parameters that contribute to the event and the features of these parameters. Parameter features can switch on and off, and vary. We need to specify how parameters change, and what the constraints are. A graph can be plotted of the change of the features of underlying events contributing to the event and show how the event occurs.

Both approaches can be used for speech modelling for synthesis (Tams 2003; Milner 1989; Wehrle and Scherer 2001).

26.1 Category Models

A category specification would involve enumerating a set of features differentiating one emotion from another in terms of inclusion or exclusion of a subset of these features. The subset could be correlated ultimately with a set of well-specified acoustic features in terms of phonetic prosodic features–fundamental frequency, formants, amplitude, duration.

For example, if *anger* has the feature of increased breathing rate, and *happiness* the feature of decreased breathing rate, and this range can be correlated with

- movement of the respiratory muscles, which are correlated with
- changes in the vocal tract, the air flow, and
- the range of acoustic features, or simply
- one feature such as duration of a segment (or syllable)

we can see a line of features moving from underlying emotion construct (*anger* as distinguishable from *happiness*) to a change in the speech waveform we can label and interpret

Developments in Speech Synthesis   Mark Tatham and Katherine Morton
© 2005 John Wiley & Sons, Ltd.   ISBN: 0-470-85538-X

as emotive content. An instantiation would produce a speech waveform on any single occasion that was recognisable by the listener as *anger* or *happiness*.

26.2 Process Models

A process specification could consist of a set of parameters that are notionally time-dependent, useful in accounting for time-varying emotive content in a conversation. As one parameter became stronger (a value on the feature increased), at any one instant, whatever the minimum time interval was, this parameter could dominate the co-occurring parameters, and ultimately be rendered in the speech waveform as an identifiable emotive effect. Process models show how a processor can transform, interpret, integrate, select and control ongoing events. A process model operates its own clock both within components and as a total system. This model is often equated with a dynamic model; but a dynamic model is not always a process model.

There is an overall process framework which may contain a number of sub-processes (equivalent to 'components'), each able to be modelled as a process. These processes can be in parallel (often the case, since process modelling comes into its own in handling parallel processing architectures), or cascaded in some hierarchical or other arrangement. Each process can have its own clock. The crude parallel in prosodics modelling would be the clocking of syntactic events (e.g. syntactically determined intonational phrases) and the clocking of rhythmic units, alignment of syntactic events by boundaries, and prosodic events.

For example, the phonological dynamic plane incorporates notional time and sequencing, but does not specify clock time. Clock time is specified in the dynamic phonetics plane which is a process. The plan of a single utterance is worked out on the phonological dynamic plane, but its time component is still notional as are constraints imposed by expression and pragmatic information channels. Real time can be assigned only during the rendering process as the result of basic procedures which take into account ongoing supervision by the 'cognitive phonetics agent'–the device's *manager*.

Part VII

Expanded Static and Dynamic Modelling

27

The Underlying Linguistics System

The constructs *dynamic* and *static modelling* were outlined briefly in Chapter 22. They are basic in the suggested approach to synthesis, providing the framework within which our proposals for high-level synthesis are made. This section is a detailed description of our use of static and dynamic models in the two domains: phonological and phonetic (Table 27.1).

The following is a description of the planes, the models and how we see the concepts as useful in underlining the contribution of phonology and phonetics to synthetic speech systems. A characterisation of the underlying expressive/emotive features we are proposing is detailed in Chapter 11 and Part IX. For a discussion on combining linguistic and expressive information, see also Part IX.

27.1 Dynamic Planes

As we use the concept, a *dynamic plane* is a working surface on which processes actually occur; this is distinct from the generalised description on *static* planes of all possible processes. The dynamic model accesses the phonological static plane for information on how to make a plan for the potential utterances. It needs an input to trigger it to draw on the static plane to process a single utterance–this input must state the requirement to produce a *particular* utterance. The process taking place on the dynamic plane needs to scan the static plane for characterisations of appropriate processes to apply at any one moment and which features/parameters are to be used in a particular instance. The rules on the dynamic plane

Table 27.1 The underlying formal linguistics system

Static plane [declarative]	Dynamic plane [procedural]
Declares an idealised general structure for the language [the grammar]	Shows how a specific structure is selected
Enumerates all conditions and options for selecting choices	Selects conditions to enable specific choices
Declares all possible occurring outputs, including exemplar outputs	Declares one specific output

Developments in Speech Synthesis   Mark Tatham and Katherine Morton
© 2005 John Wiley & Sons, Ltd.   ISBN: 0-470-85538-X

specify action. What to retrieve from the static plane is determined by the input. For example, if the instruction is '*You need a* [t] *here*', a rule on the dynamic plane enables the system to go and fetch the specification for a [t] from the static plane for use in this particular instantiation.

Notional time from the static plane can be called to and modified on the dynamic plane. For example, the lexicon may specify a *long vowel*, and a dialect variation may be to relatively (notionally) shorten this vowel. On the phonological dynamic plane this information has been called and used to produce an appropriate phonological plan which includes the shortened vowel. The dynamic phonetic plane takes in this plan in order to render it. An important part of rendering involves realising notional time as real time, so the 'shortened' vowel is now assigned a clock time duration. The consequences of doing this are predicted and made available to the 'cognitive phonetic agent' (CPA), which does something about the situation if it generates or is likely to generate an error.

A plan for a single instantiation has been generated on the phonological dynamic plane and is one of the possibilities enumerated on the phonological static plane. For the rendering process the dynamic phonetics needs:

1 the plan
2 information about phonetic processes declared on the phonetic static plane–it calls the appropriate information in accordance with the plan.

27.2 Computational Dynamic Phonology for Synthesis

The input to the '*dynamic phonology*' is the output of the syntactic component consisting of a well-ordered string of words (morphemes) derived from the lexicon, and including abstract underlying phonological features such as length, voicing, syllable etc.

The output of the dynamic phonology is handed over in turn to the dynamic phonetics. Using terms common in the early days of transformational grammar, this sequencing can be seen as like a competence and performance characterisation–the static model reflects competence, the knowledge needed; the dynamic model describes aspects of performance, what is being done and actually produced.

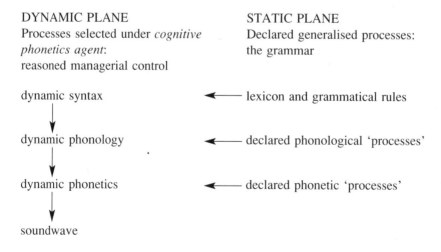

DYNAMIC PLANE   STATIC PLANE
Processes selected under *cognitive*   Declared generalised processes:
phonetics agent: the grammar
reasoned managerial control

dynamic syntax   ◀——— lexicon and grammatical rules

dynamic phonology   ◀——— declared phonological 'processes'

dynamic phonetics   ◀——— declared phonetic 'processes'

soundwave

In many cases the plan may be equivalent to an exemplar sequence of phonological units and rules computed on the *static phonological plane*, but technically this will be a coincidence.

One feature important for our proposal is *prosody* and how it is handled in phonology. The prosody, we suggest, operates as a framework for linguistic and cognitive expressive content. *Plain* (sometimes referred to as 'neutral') linguistic and cognitive expressive content have independent abstract existence, so they need independent characterisation. The general prosodic structure also needs separate characterisation because it exists independently of its ability to host linguistic and expressive content. The independent characterisation shows generalised prosodic *structure* and reveals the 'hooks' by which linguistic and expressive content enter into the structure to produce a composite prosody.

The overall architecture may look like this:

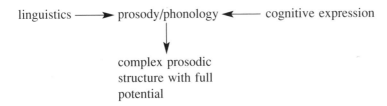

Although expression can be independently characterised, in reality it does not exist outside the person. The existence of expression can be analogised with modulating a carrier. In this analogy, there can be no modulation without a carrier, and also the carrier itself does not communicate the information (information is conveyed by the way it is modulated). Another analogy could be painting–there cannot be a painting without a canvas which supports and provides a continuity of surface for the painting, but it is (usually) not seen directly.

Dynamic phonology can be thought of as the phonological blackboard where the phonological plan for a particular sentence utterance is worked out. Specific dynamic processes are involved and these operate on specific generalised processes taken from among those declared in static phonology.

27.3 Computational Dynamic Phonetics for Synthesis

Dynamic phonetics receives the plan from dynamic phonology, and proceeds to turn it into an utterance. The dynamic phonetic plane is responsible for the physical rendering. It takes notional time (e.g. [+length]) from the phonological input, and re-represents it as clock time, incorporating constraints called from the phonetic static plane. It specifies segment sequence management highlighting coarticulatory phenomena which are clock-time constrained. Dynamic procedures call routines from static phonetics to handle segment sequences for this particular utterance. The final specification is a string of units or (in a more advanced phonetics) a set of paralleled gestures needed for input to the motor control system that moves the articulators into the correct vocal tract shape to produce the intended utterance. This plane draws on what is physically and perceptually possible to do, as described within the 'phonetic static plane', and uses this information to describes the actual action needed.

27.4 Adding *How*, *What* and Notions of Time

We use the terms *how* and *what* as follows:

- *What* can happen is described in the abstract for all utterances on the static plane, and *what* does happen is also described in concrete terms for a single instantiation on the dynamic plane.
- *How* is specified notionally on the static plane as a wide range of possibilities from which the operations on the dynamic plane must select. The selection is made according to various conditions that are current (occurring now, at this moment), specific constraints and environmental requirements etc. Principal among these is an idea of how the timing of this utterance is to be handled–is this part of the speech to be fast, or slow, or whatever?–and the consequences of this are now to be computed in the rendering process. The basic structure of *how* is therefore already available in general on the static plane.

What and *how* appear on both planes, but in different contexts–in the context of all possibilities on the static plane, and in the context of *this* possibility on the dynamic plane. An example from syntax: *how* all sentences are eventually, on the surface, made up of well-ordered linear sequences of words is specified on the static plane. For any one sentence, how the static generalisations are used in the rendering process to produce the *how* of a single instantiation is computed on the dynamic syntactic plane.

Thus dynamic and static models can show both what happens and how it happens. Dynamic models can include the idea of time when appropriate, doing so, for example, in dynamic phonetics when rendering a specific utterance.

Constituents of the phenomenon that potentially move through real time can be specified as notional time in a dynamic model. Real time can be specified when instantiated by procedures called into play during the instantiation. An example might be the way changes are sometimes needed in the motor control system when speaking. Muscle movement is almost always needed *prior* to the occurrence of an actual sound (we know this from experiments involving electromyography). That is, muscle activity (e.g. for the lip closure associated with the stops [b, p]) begins measurably before the appearance of features in the acoustic waveshape perceived by the listener (Tatham and Morton 1969).

27.5 Static Planes

As we use the concept, these planes hold sets of static descriptions or declarations about potential utterances. The descriptions include features/parameters, rules and procedures. They can be in the form of a list, or a dictionary with an object and a look-up table of features that characterise that object. They can consist of a simple sets of information (data, facts), or a structured knowledge base (information, its internal relationships, and how to use it).

Consider, for example, the table of phone-sized objects in a typical formant synthesis text-to-speech system: the JSRU system (Holmes *et al.* 1964). Here, we find abstract parametric representations of each potential segment, with values specified for each parameter, along with a notional or 'intrinsic' duration. The information is static–it is there to be used, but it has not yet *been* used–but most importantly it is exhaustive of the abstract possibilities. The form taken in this system is interesting: there is a highly generalised representation specifying the set of all possible parameters for all possible phone segments, together with a statement of all possible values (in fact, the ranges) for values for each parameter. A less abstract form follows–a repetition of this highly generalised for representation, but now particularised for all possible representations of any one phone. This is a kind of 'average' of what might be expected from all possible potential instantiations (called by Holmes and others 'targets'). How these phone declarations–the definition of what a phone is in the abstract and of what particular phones might be in the abstract–is added to generalised statements of how they might be used to form the complete structured dataset.

Features in a static description are not time-varying, although the information can be given that these features/parameters may be called upon by a procedure that requires this specification in a time-varying context. Once again a useful example is the single intrinsic duration parameter for phones in the JSRU text-to-speech system. This procedure will be called from the dynamic plane in order to produce a specific event within in the context of clock-time constraints. That is, notional time is declared in the static model, and will be given specific values later when the information is accessed to produce a one-off event. Notional time generally describes phenomena such as 'length' and sequencing.

27.6 Computational Static Phonology for Synthesis

The phonological static plane includes a characterisation or declaration of cognitive segmental processes. It provides a characterisation of what linguists call the *sound patterning* of a particular language. Some of the statements and rules about the patterning can be found in many if not all languages, and can be said to describe universal tendencies. For example, syllables often begin phonologically with progressively increasing sonority and end with progressively decreasing sonority. Much work has been done in establishing language universals (Comrie 1989; Kirby 1999). However, even though cross-language studies show that researchers can formulate similar descriptions, this does not mean that there is a commonality that is useful for synthesis purposes. And the generally agreed position is that phonology describes language-specific phenomena. We suggest that speech researchers might usefully adopt this position because we are interested in simulating a particular language output. The only exception we might want to make is when considering the addition of an extra tier in the static phonology to account for the derivation of different accent forms from the abstract, single, overarching language (see Chapter 13). Consider a couple of examples of particular commonly cited rules of the phonology:

- In English, a language-specific rule is: if there are three consonants at the beginning of a syllable, the first of these must be /s/. Another rule, specific to English, is that an underlying abstract segment /L/ is phonologically processed to a palatalized /l/ before vowels but a velarised /l/ after vowels or before syllable final consonants–that is, at the beginning and end of syllables respectively. This is a generalisation within English phonology; whether other languages can be described as planned in a similar way is not relevant to simulating English.
- *Length* may be a feature, but no specification is given in phonology as to how long or short length is in clock time. These values are stated in the static phonetics model which specifies the possible durations and physical constraints on the ability of the speaker to

produce the intended sound and the ability of the listener to perceive it. However, the phonology of English (and many other languages) includes the rule that vowels lengthen in syllables with a coda which includes a [+voice] consonant–a declaration which will need rendering in physical world correlates by the static phonetics (for all possible utterances) and the dynamic phonetics (for one specific utterance).

In terms of the plan, static phonology is responsible for declaring the generalised phonological structure of all possible sentences–it provides the sound shape interpretation of *all* sentences which can be generated by the syntax/semantics. Exemplar sentence plans are possible; their function is to give illustrations of how particular subsets of the processes operate.

As a further example of what is to be found on the static phonological plane, we might move away from segmental processes and consider non-expressive stressing at the word level to achieve contrast between syntactic categories, or at the phrase level to achieve contrastive emphasis. Eventual phonetic rendering of these will use varying amounts of amplitude change and fundamental frequency change beyond what is expected–the prominence is coded in the acoustic signal by the predicted varying values of particular parameters. Examples are:

- cont<u>rast</u> (verb) as opposed to <u>cont</u>rast (noun)
- per<u>mit</u> (verb) as opposed to <u>per</u>mit (noun)
- 'I said press the <u>control</u> key, not the <u>shift</u> key.'
- 'He's from <u>eastern</u> Canada, not <u>western</u> Canada.'

This type of variation is not expressive; it is described by linguistics for a specific language, and may be rendered differently acoustically in different languages or even accents of the same language. In addition, these words or phases can also take emotive content; so the two are dissociated. Clearly the implementation of phonological stress is essential in synthesis systems, and is already available in most; but it is important to keep this type of stressing distinct from non-linguistic stressing.

Thus the utterance plan is the final outcome of phonological processing. It is produced on the dynamic plane after accessing the knowledge found on the static plane. Static phonology provides the units and processing for planning of all possible utterances. It provides the sound shape interpretation of all sentences which can be generated by the syntax and semantics. The exemplar sentences produced can illustrate the possibilities in an abstract form; these exemplar sentences are not actual sentences.

27.7 The Term *Process* in Linguistics

We should point out a potential confusion arising from the use of the term phonological *process* in linguistics. It does not mean a process involving clock time. Similarly in syntax, a tree diagram is a graphical representation of a set of processes, and the rules of syntax and phonology are also called *processes*. In fact these processes and their algebraic or graphical representations are characterisations of data structures. Because the data structures are arranged hierarchically, linguists regard dropping a level as a *process* within the overall data structure. The data of linguistics derives mainly from observations by the linguist of the output of actual language production or from reports of native speaker intuitions, and the generalised architecture of linguistics is a hierarchical arrangement of cascaded 'rewrites' of data structures. It is important not to confuse the use of the term *process* in linguistics with its use in, say, computer science or elsewhere in this book.

In linguistics it is considered legitimate to write rules describing processes which give rise to an output. Since the individual utterances output from the system–the *logically* final data structures–constitute a subset of all possible instantiations of the generalised data structure, the value of this kind of data as a base for the complete generalised data structure is in some question. What we mean here is that a main source of data for linguistics comprises utterances which people have spoken or written–instances emanating from the dynamic plane. The backwards transition from this data to a full characterisation of what is on the static plane (the focus of theoretical linguistics) is held by some to be an inexplicit procedure within the science. There are two points here:

- How linguists arrive at a characterisation of the grammar of a language is not relevant to speech synthesis researchers. However, the general properties of the grammar and what it actually *is*, and what it can be legitimately used for, *are* relevant.
- Taking the linguists' grammar and 'running' it is not a legitimate procedure. What this produces is an exemplar instantiation from the set of all possible outputs–technically the 'language'. Running a set of logically descriptive data structures as though they were time-governed computational processes is not legitimate. This point is explained more fully in Tatham and Morton (2003). An example of the consequences of such a technical error would be the production of synthetic speech with a neutral prosodics arising from the fact that in linguistics phonology legitimately handles no more than such neutral prosodics. It is a mistake to believe that phonology has therefore described adequately what speech synthesis experts might actually need–an extended phonology which is very much more than the potential 'carrier' of expressive content. Remember, this neutral phonology is an abstraction which *cannot* exist in the physical world of synthesis.

In traditional theoretical phonology, the domain of the science–termed by us the static phonological plane–characterises neutral linguistic processes, reflecting the linguistic features of a sentence and interpreting these within the sentence domain as an abstract 'sound shape'. The phonological meaning derives only from the available semantic information and does not normally include explicitly expressive or pragmatic information. We suggest, however, that this latter information is essential for natural speech. The descriptions relate to an exemplar derivation, not an actual sentence or utterance. In most cases, it will probably be the case that the plan for an actual utterance, based on the requirement to speak an actual sentence, is the same as an exemplar plan given in static phonology. One difference may be that the plan for an actual sentence includes, or highlights, previously included hooks for linguistic expressive content or pragmatic effects.

There are two possible ways of adopting these ideas, but in our view only one of these will be productive. The first model is the one we do *not* favour, and potentially is a misuse of linguistics:

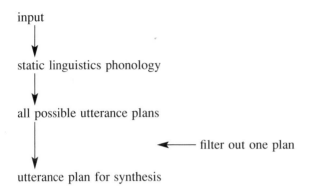

The second model is better and uses linguistics to supply information throughout the dynamic derivation of a single utterance plan. Any exemplar plans which the static phonology might 'produce' are incidental:

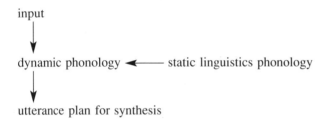

One advantage of drawing on static characterisations for continuously feeding dynamic processes is that external information (pragmatic etc.) can be introduced legitimately as part of the dynamic process. Static generative phonology makes no formal provision for doing this that might be used for our purposes.

27.8 Computational Static Phonetics for Synthesis

The *principles* of phonetic rendering needed for turning all sentences characterised by the static phonology into utterances may in general be universal (see below) and relate to matters such as the correlation between phonological features and phonetic features. For example:

1 The correlation between phonological voicing and how this is potentially rendered using a periodic source needs to be made explicit.
2 The constraints of phonetic rendering affect what is and what is not possible as part of phonological planning. These include the relationship between articulatory shapes and the soundwaves they underlie, the constraints of the mechanics of the articulatory system and its control, the mechanism available for controlling the movements of the speech organs, etc.
3 Timing constraints will need enumerating here, because although they take in phenomena such as coarticulation during the course of the rendering of an utterance, these are general principles which can be characterised independently of any one actual utterance.

4 For any one sentence characterised as an exemplar derivation in the phonology, it is possible to provide an exemplar phonetic rendering (on the static phonetic plane) which shows what such a sentence, one among all sentences, would look like if it were an utterance a speaker wanted to produce.

We must be careful about the concept of universality here. There are a great many universal principles underlying phonetic rendering, most of which are down to external factors such as motor control, mechanics, aerodynamics, acoustics etc. which have nothing to do in themselves with linguistics (i.e. they are independently modelled in some other discipline). But because of the possibility of language-specific cognitive intervention in the rendering process, it may appear that *on the surface* much is language-specific. This is what lies behind much of the *theory of cognitive phonetics* (Tatham 1986a,b, 1995) which introduced the notion of cognitive intervention from biopsychology to phonetics. Universally governed phenomena can be manipulated through cognitive intervention, and the whole process carefully managed by the cognitive phonetics agent.

Time-governed constraints are enumerated here, and the dynamic phonetics calls on these. This plane specifies general processes needed to render the segments, syllables and phonological prosodic features. In fact all the detail noted for the physical production of speech is stated on this plane–information such as the interrelation of air pressure and vocal fold vibration, phonological nasalisation and velum opening/closing, stop consonants and degree of vocal cord vibration (including absence), coarticulation constraints and limits, and very importantly the degree to which some physical constraints present in the system at rest can be overcome, and so on.

For example, declination is a feature of some languages, to varying degrees. The amount of declination in English is sometimes reported as greater than that in French. It is thought that the English declination represents the system more nearly minimally functioning, whereas an effort is needed to overcome the state of the system at rest in French. An effort is needed to maintain the air pressure until the end of the utterance to achieve the perceptual effect of 'flat intonation'.

Because of the possibility of cognitive intervention it is necessary to account within phonetics for principles which on the face of it are not to do with language. Early transformational linguists (Chomsky and Halle 1968, Chapter 7) dismissed phonetics from linguistics on this basis (Keating 1990). But since we now know about cognitive intervention as a general somatic principle, which as speakers *we are not able to negate*, it becomes necessary to include a model of what it is that is being modified.

When the dynamic phonetics has taken the sketch, or outline, of the utterance–the plan–and filled in the details as part of the rendering process, the output is taken by the motor control system and the speaker produces the utterances which have encoded all the syntactic, pragmatic, semantic, phonological and dialect information etc; and *in addition* modifications added by the physical characteristics of the individual's vocal tract (such as large resonating chambers, thick lips, long vocal tract, small sinus chambers). The result is the speech waveform, expressing the thoughts, expressive/emotive states and the individual characteristics of the speaker.

27.9 Supervision

Speaking is an ongoing event and a fixed descriptive system cannot, by definition, adapt to changing environments and general context (Morton 1992). The construct 'agent' is used (Garland and Alterman 2004; Luck *et al.* 2004). We interpret this for our purposes as a *cognitive phonetics agent* (Tatham 1995; Tatham and Morton 2003), or CPA. This is a cognitive object able to monitor, reason about its findings and issue modifications of the speech output, depending on changes in context, including the behaviour of listeners. It controls, organizes, takes into account the limits of what can be done, instructs the system to modify constraints on the speech production mechanism, to integrate with other information such as feedback from the listener's reaction to speech, or to the speaker's reaction to the listener's reaction, in a conversation.

The information channels which call for specific phonological or phonetic effects have to be timed for and associated with each utterance; failure to do so would destroy the coherence and integrity of a speaker's output. The immediate requirement can be met by constantly renewing requests to the dynamic phonology (and later the dynamic phonetics) to reselect from among the choices possible listed on the corresponding static planes. Such renewed requests are, in our model, under the direction of the CPA.

The CPA monitors the way phonological prosodic information is included in the plan, and also how phonetic prosodic rendering is carried out. We propose that expression/ emotion are encoded into the prosodics, the cognitive expressive content into phonological prosodics, and the basic biological emotion manifest in the state of the physical system. The separation of cognitive and physical aspects here is significant and is compatible, we believe, with current thinking in biopsychology. The CPA will monitor the physical system, and see that the plan is carried out appropriately by phonetic rendering. The CPA can also monitor how a generalised perceptual system might deal with the putative output, with a view to optimising the equation between production and perception. Ideally the speech units intended are equivalent to the units perceived. The CPA will also track actual perception by observing the behaviour of listeners. There are bound to be imperfections in phonetic rendering, but many of these will be repaired by error-correction procedures which are part of the perceptual process. The CPA will predict error correction, and observe when it fails, backtracking to improve rendering quality.

27.10 Time Constraints

Time constraints in the rendering process may introduce errors. This possibility is, we propose, continuously monitored by the CPA to make sure that rendering is performed as well as is possible–or as well as is necessary (see the above mention of error correction) according to the phonological plan, but also according to known constraints of the perceptual processes which have to deal with the actual utterance. Speaking fast, for example, can change coarticulatory effects to the point where it is predicted that perception will fail. The CPA can monitor this and slow down speech rate, or increase precision of articulation, or both accordingly. There may be a sudden requirement for temporary or *ad hoc* and immediate changes to pre-calculated expressive content; the CPA is sensitive to clock time, and can modify 'target' values derived in the static plane. Phonetic prosodic effects–the rendering of the phonological prosody–can be regulated by the CPA to produce appropriate changes during a conversation; either enhancement or suppression. These are highly complex issues in modelling speech production. But we include them here in some detail precisely because they bear so intimately on the very basis of naturalness in speech–the inclusion of expression and the continuous sensitivity to the possibility of error.

During phonetic rendering, alternatives present themselves, although phonetic constraints may be strong. For example, there are a number of mechanical constraints which operate on certain sequences of segments–in a more traditional model, the so-called 'coarticulatory constraints'. Consider, for example, the way in which the tongue blade moves from a forward position of contact with the alveolar ridge to a more retracted position during rendering of the phonological sequence /ta/ (orthographic *tar*). The coarticulatory processes are characterised in general terms in static phonetics, and called on by dynamic phonetics in any one utterance. However, the general terms do not tell us whether the utterance is actually going to be spoken quickly or slowly. One function of the CPA is to tweak dynamic phonetic processes so they occur as near optimally as possible or perhaps necessary. Although phonetic processes seen simply as physical processes are generally considered to be universal, the intervention of the CPA modifies them in a way which is sensitive to various factors:

- the requirements of a particular language or dialect–for example the 'extra' nasalisation of inter-nasal oral vowels in some accents of English, or the precision of articulation needed when the articulatory space is crowded as in the comparative precision afforded [s] and [ʃ] in English compared with [s] in Spanish and Greek);
- an immediate requirement for expressive rendering which may involve increased precision, say, for detailed rendering of an emphasized word or phrase;
- perceptual difficulties noted in the listener–for example a noisy environment (short term) or slight deafness (long term) which requires greater precision of articulation perhaps 'on demand'.

Thus, the CPA is not seen as a component or as a level but a supervising or managing *agent*. The plans are provided by phonology, phonetics renders the plans, and the CPA ensures that all components do their job appropriately to make everything come together to provide an adequate signal for the listener. The CPA is sensitive to user needs; it is in constant touch with a predictive perceptual model and the various forms of feedback available for its use.

27.11 Summary of the Phonological and Phonetic Models

The production of an utterance soundwave requires a minimum of data and processes. Without consideration of expressive content, these include:

1 a *sentence* (the basic domain for linguistic processing–though not the only one) for inputting to dynamic phonology which references a static phonological database of available units and processes;
2 a *phonological plan* to characterise a representation of sound shape and basic prosodics suitable for phonetic rendering to produce a waveform;
3 a *dynamic phonetics* to render the plan by addressing a phonetic database residing in static phonetics;

4 a *CPA* to supervise the relevant inputs to each module and how they are processed to produce an appropriate output.

In the next chapter, which discusses expression/emotion for inclusion in utterances, we will add what is needed for an utterance with expressive/emotive content:

1 a *sentence* for phonological processing–stated as a linguistic characterisation of syntactic, semantic and pragmatic information, together with cognitively sourced expressive/emotive information, at this stage indicated by choice of words and grammatical constructions;
2 a *phonological plan* upon which phonetics can operate to produce a series of articulations (better: articulatory gestures) which when enacted produce a speech waveform (added to the plan must be some cognitively sourced expressive/emotive information; this gives us prosodically encoded expressive content, rather than the syntactically encoded content of (1) above;
3 a *dynamic phonetics* which renders the plan within the constraints of the biological system (specified as part of static phonetics, together with other types of constraint–mechanical etc.);
4 a sufficient specification of the *biologically based modifications* which occur when the overall biological stance changes during an emotion state, and how that stance further changes as time unfolds.

28

Planes for Synthesis

It is on the dynamic planes that processes involved in developing particular utterances occur. Central to the plane concept is their 'blackboard areas' or working spaces on which the following are to be found:

1 a *sentence* for phonological interpretation (output from dynamic sentence processing, input to dynamic phonology);
2 a *phonological plan* (output from dynamic phonology, input to dynamic phonetics);
3 a set of *phonetic instructions* to render the *specific* plan (called to dynamic phonetics from static phonetics);
4 necessary *pragmatic information* located by the CPA and feeding into the dynamic phonological and planes–this is the cognitive source of expressive content, and is additionally supervised by the CPA;
5 *biological information* providing the additional source of expressive or emotive content, also supervised by the CPA and feeding into the dynamic phonetic plane.

The CPA supervises the processes in (2) and (3) to produce an utterance; it monitors (4) and (5) to modify this process. Notice that under (4), pragmatic information (which is cognitively sourced) feeds both dynamic phonological and phonetic planes. This is because although dynamic phonetics is primarily about physical processes it nevertheless permits cognitive intervention–and some of that intervention is from pragmatics. Under (5), biological information can feed only to phonetics rather than to phonology because it has no cognitive content.

> Note the use of terminology. The abstract phonological prosodic framework is about *intonation*, *stress* and *rhythm*, whereas the physical phonetic prosodic framework is about *fundamental frequency* (often used in rendering intonation), *amplitude* and *local durational variation* (rendering stress), and *rhythmic timing* (often rendering rhythm). The correlations between phonological and phonetic features are *nonlinear* and *non-exclusive*.

Speech communicates the underlying plain message and expression which envelops it simultaneously–both aspects of the communication are *always* present. The results of the

---

Developments in Speech Synthesis   Mark Tatham and Katherine Morton
© 2005 John Wiley & Sons, Ltd.   ISBN: 0-470-85538-X

cognitive phonological system are rendered by the physical phonetic system which itself has characteristics that interpret and add information to the intended message. Both message and expression convey information about the speaker–for example, information about the speaker's attitude, belief and emotion state at the time of speaking. The message may also be intended to evoke a particular attitude, belief or emotional state *in the listener*.

Since speech researchers regard the waveform as divisible into two parts, the message and emotive content, we can model the processes as separate; but we must remember that they are separate only in the model, not in the speaker or the listener. We suggest an approach which does not separate these two aspects of the *waveform*, but recognises the advantages of modelling two separate *underlying* processes: cognitive and biological.

We shall focus on the final stages in spoken language production and look at modelling production in general terms and also speech produced at any one instant. The descriptions are of the intended sound pattern of the language (characterised phonologically) and the vocal tract configuration and aerodynamics (characterised phonetically). Short-term events, as well as longer term phenomena that continue over short-term articulations of segments, will be described.

The model we are proposing attempts to characterise human speech production in such a way that it has relevance to the field of speech synthesis–fundamentally a computational simulation of the human process. We are aware of gaps, just as we are aware of some areas of departure from conventional ideas in linguistics; but we are even more aware of the overwhelming need for coherence and computability. We are confident that adhering to the latter will enable gaps to be plugged as and when detected, and transparent changes in the theory to be incorporated as more data emerges. But for the moment our concern is to do what we can to enable a modest moving forward of work in synthesis, particularly in the very difficult area of expressive naturalness.

Part VIII

The Prosodic Framework, Coding and Intonation

29

The Phonological Prosodic Framework

When we consider speaking as a modality of expression, the framework we outline is called *phonological prosody*. This refers to an overarching set of features that include intonation and duration. Elements in the prosody are the IP (intonational phrase), the AG (accent group), rhythmic unit, and the syllable. These terms are explained at the end of this chapter. The phonological prosody contains expressive/emotive information that is cognitively sourced; the *biologically* sourced information is part of the phonetics and is specified at a later stage. The expression wrapper is seen as a context within which the phonological prosodic data structure operates.

The outline of the model proposed above arranges the elements hierarchically–that is, the higher levels contain within them the lower levels. Unless otherwise specified, features or attributes of the higher levels will also be found on the lower levels, but features or attributes specified on lower levels do not migrate upwards.

An item enclosed in brackets–for example, <syllable>–is an element. There will be attributes on these elements for specifying detail (i.e. giving value to attribute parameters). For example, <IP> might take an attribute contour which itself could have values such as 'rising', 'falling', 'level', 'discontinuous'–even a more detailed value generalised as 'shape of curve' allowing for the possibility of a number of different pre-formed curve shapes. Shape of the curve is essential for instantiation, since the shape will determine the overall pattern for the fundamental frequency. These patterns are usually characterised as abstract iconic graphs; very little work has been done on classifying intonation which is capable of being directly transferred to a contour suitable for subtle synthesis. A useful interim system, based on Pierrehumbert (1981), is to assign abstract markers on syllables which indicate 'high' and 'low' points in the intonation contour as characterised on the static phonological plane. A way of indicating actual but stylised physical contours has been introduced into the SSML markup conventions by the W3C working group (see Chapter 18)–once again relying on interpolation between a few pivot points in the physical contour.

Figure 29.1 shows an example of basic linguistic intonation assignment using the method developed for the SPRUCE high-level synthesis system. Boundary markers are introduced into a sentence using an algorithm based on a syntactic parse:

Developments in Speech Synthesis   Mark Tatham and Katherine Morton
© 2005 John Wiley & Sons, Ltd.   ISBN: 0-470-85538-X

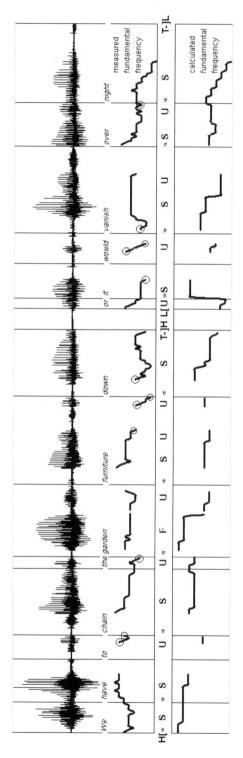

Figure 29.1 The utterance is *We have to chain the garden furniture down or it would vanish overnight*. At the top is a waveform of human speech with, immediately below, the measured fundamental frequency curve. Below this is the abstract symbolic phonological marking assigned to this sentence by the SPRUCE algorithm. At the bottom of the figure is the fundamental frequency rendering of the high-level markup. It is clear where the rendering algorithm has delivered a curve which does not match the human example–though it is not necessarily the case that the calculated version cannot stand in its own right: human beings will render the *same* markup differently. The measured fundamental frequency curve also has examples of micro-intonation circled (see also Figure 15.1). Microprosody of this kind is not part of linguistic prosody and does not therefore figure in the SPRUCE prosody algorithms. It is acquired directly from the database when the system drives a unit concatenation rendering, or is calculated as part of coarticulation when the system drives a parametric target-based rendering. This figure has been adapted from Tatham *et al.* (2000).

1 High and low markers are assigned.
2 A specially designed algorithm converts the abstract markup into putative fundamental frequency curves.
3 A custom procedure is applied for introducing a 'wobble' into the assigned values to simulate.
4 A waveform is synthesised either by formant synthesis or by appropriately modifying concatenated units called from a database.

In XML, attributes can have relative values or specific integer (or other) values. Thus:

- <syllable>

  - *Attribute:* duration ["long", "mid", "short"] or [integer, range = "10–450", default = "250"]
  - *Attribute:* peak amplitude ["high", "mid", "low"] or [integer, range = "–35 to 5", default = "–15")

Here the element <syllable> can take a duration attribute with either relative, abstract values (e.g. for example, "long", "short") or can, to ease integrating with phonetic prosody, have a specific integer value (30, 184, etc.) in milliseconds, with range (10–450ms) and default value (250 ms) made explicit. Similarly a peak-amplitude attribute can be introduced with a relative range (e.g. "high", "mid", "low"] or can have a specific integer value (–25 or –10) in decibels (dB), with a range of –35 to –5dB and a default value of –15dB. A similar approach can be taken with individual phone-sized elements if necessary.

What is important explicitly here (and implicitly in many of the SSML attributes) is the way the proposed approach can be made to bridge the abstract/physical gap. For example, "long" and "low" are attributes which can be used of both length and loudness (perceived qualities) as well as duration and amplitude (quantifiable properties). In the end, though, careful and explicit use of attributes (and also in the choice of the elements themselves) in this way is no more than an ad hoc approach to solving the abstract/physical relationship problem. In the linguistics area the problem is encountered all the time at the phonology–phonetics interface, and in psychology at the biology–cognition interface.

29.1 Characterising the Phonological and Phonetic Planes

This section is an outline of a possible way to characterise the static phonological and phonetic planes using XML notation. Following on from procedures for accessing the static planes from the phonological and phonetic dynamic planes, we give an outline of these procedures in pseudo-code. Directive and polite expression, as well angry emotion, are set out in the fragments of code below.

Consider the utterance *Please hold the line* as spoken by a call centre voice. This sentence has arrived on the dynamic phonological plane, at the highest intonational tier, and is assigned the data structure element <IP>–the widest intonational domain. Within this domain there are three sequenced (AND-ed) accent groups, <AG> elements: the first of these contains a single stressed syllable (stress="1"), the second a focus (Artstein 2004) stressed syllable (stress="2") followed by an unstressed syllable (stress="0"), and the third a single stressed syllable. The sentence is initially assigned the general case declarative markup:

```
<expression>
      <phonological_prosody>
            <IP>
                  <AG>
                        <syllable stress="1"> please </syllable>
                  </AG>
                  <AG>
                        <syllable stress="2"> hold </syllable>
                        <syllable stress="0"> the </syllable>
                  </AG>
                  <AG>
                        <syllable stress="1"> line </syllable>
                  </AG>
            </IP>
      </phonological_prosody>
</expression>
```

where

- <IP> is an intonational phrase: the domain of an intonation contour
- <AG> is an accent group: a small intonation unit within the <IP>
- <syllable> is the lowest unit (node) of prosodic phonology: phonology within a prosodic framework, itself within the expression framework
- <syllable stress="0"> is an unstressed syllable
- <syllable stress="1"> is a syllable with primary stress
- <syllable stress="2"> is a nuclear stressed syllable signalling focus.

Another symbol sometimes needed, but not illustrated in this example, is $, a null syllable. All rhythmic units must begin with a stressed syllable in this system, and to satisfy this condition if one is not present in the utterance the null syllable is inserted to indicate an empty stressed syllable. For example, the utterance *The house is white* is given as:

| $ *The* | *house is* | *white* |, with <AG> boundaries indicated by |.

The derivation, the output utterance plan from the dynamic phonological plane, would look as follows with syllable detail added:

```
<expression>
      <phonological_prosody>
            <IP>
                  <AG>
                        <syllable stress="1">
                              <onset> pl </onset>
                              <rhyme>
                                    <nucleus> i </nucleus>
                                    <coda> z </coda>
```

```
                                              </rhyme>
                                    </syllable>
                          </AG>
                          <AG>
                                    <syllable stress="2">
                                              <onset>  h  </onset>
                                              <rhyme>
                                                        <nucleus>  ɔ  </nucleus>
                                                        <coda>  ld  </coda>
                                              </rhyme>
                                    </syllable>
                                    <syllable stress="0">
                                              <onset>  ð  </onset>
                                              <rhyme>
                                                        <nucleus>  ə  </nucleus>
                                                        <coda  >  0  </coda>
                                              </rhyme>
                                    </syllable>
                          </AG>
                          <AG>
                                    <syllable stress="1">
                                              <onset>  l  </onset>
                                              <rhyme>
                                                        <nucleus>  ai  </nucleus>
                                                        <coda>  n  </coda>
                                              </rhyme>
                                    </syllable>
                          </AG>
                </IP>
          </phonological_prosody>
</expression>
```

where a 0 in the coda element of *the* indicates a null coda–the syllable ends with the nucleus.

<expression> is an element wrapping the entire utterance, including the output of the logically prior syntactic component. Here, though, we are concerned only with possible phonological prosodic exponents such as pausing, and other prosodic features that are rendered by phonetic prosody.

A default intonation contour (though not a 'neutral' one) characterised on the static phonological plane is retrieved by dynamic phonological processes on to the dynamic phonological plane. This then proceeds to the dynamic phonetics which retrieves phonetic information from the static phonetic plane and calculates the numerical values which will trigger the appropriate listener response. Looking at another example, rhythm, the procedure can be characterised as:

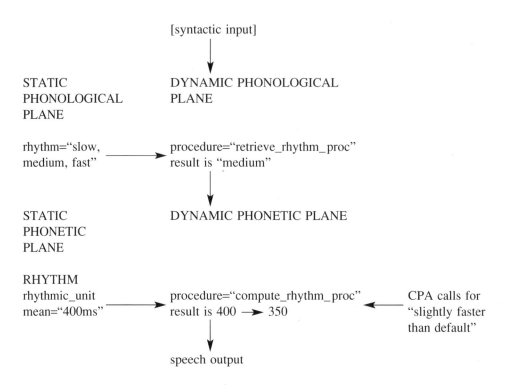

XML can declare procedures as an attribute (this is the technical means of including procedures within the declaration of data structures); for example:

- <phonological_element procedure="retrieve_abstract_value">–[PLANNING]
- <phonetic_element procedure="compute_numerical_value">–[RENDERING]

Thus the kind of the effect that can be characterised in this way could be

- intended effects, such as a pause of specific duration after 'please' in *Please hold the line* in order to convey a directive rather than an instruction;
- unintended effects, such as a relaxing of the vocal tract initiated by the CPA, thus conveying a more relaxed attitude.

So, this is a suggested mechanism for incorporating some expressive features during utterance rendering. In the example given, expressive content has been included in the final signal by having the CPA intervene in the phonetic rendering process–which in turn has made a principled selection from the static possibilities.

The detection of emotive effects can be triggered in the listener by varying the rate of the utterance (overall number of rhythmic units per unit time) which in turn will affect coarticulation at the segmental level. Timing relations will be changed–the rate of delivery of individual words within the utterance–by interrupting the rhythm by, for example, the insertion pauses. One problem will be that the effect of changing syllable durations may enlist a self-adjusting system designed to keep rhythmic units relatively constant; and this may need overriding for emphatic effect. Thus the characterisation must be hierarchical and include reference to this nonlinear relationship. For example, take two renderings of the utterance *I was in Scotland last September . . .* :

plain, with no emphasis: | *I was in* | *Scot.land* | *last Sep* | *tem.ber* || . . .
emphasis: | *I was in* | *Scot.land* | *last Sep* | *TEM.ber* || *not Oc* | *to.ber*

The calculated duration of the rhythmic unit | *tem.ber* ||, if held constant for the second sentence where *TEM* is increased in duration to convey emphasis, will require a shortening of the unstressed syllable *ber*. However, this is not what can often be observed in data from human beings: often the *ber* syllable will be the same duration in both renderings and the rhythmic unit duration adjusted to take account of the increased duration of the stressed syllable. On the face of it this seems a simple matter to handle, but the fact is that in human speech rhythmic unit rate is observed on the surface to be continuously variable and it is no small matter to decide exactly how to incorporate this. Our own approach is, as we show, to have an abstract default which translates into a physical default which in turn is continuously adjusted in line with the consequences of CPA intervention. Thus an observer would not find a constant rhythmic unit duration in the soundwave without a multi-layered analysis. Our model, rather simplistically exemplified here, explains in principle the conclusion that isochrony in speech–perceived equidistance in time between stressed syllables–can exist for the speaker in the utterance plan and in the utterance plan assigned to the signal by the listener, *while not being measurable in the signal.*

Definitions used for rhythm data structures and procedures:

- *Rate*–the number of syllables or rhythmic units, or words, in a given unit of time. Different researchers use different units here, but we use rhythmic units to calculate rate.
- *Accent group* (AG) –a sequence of one or more feet: a unit of intonation.
- *Foot*–an abstract unit (phonology) of rhythm, consisting minimally of one stressed syllable optionally followed by up to, say, four unstressed syllables. Feet can span word boundaries.
- *Rhythmic unit*–a physical unit (phonetics) of rhythm: the time from one stressed syllable to the next (the physical equivalent of the abstract foot).
- *Rhythm*–the patterning of rhythmic units within rate.
- *Isochrony*–perceived equidistance in time from one stressed syllable to the next.

30

Sample Code

In human speech there are two types of expressive content:

- cognitively sourced
- biologically sourced.

The examples of expression used here are both cognitively sourced: 'directive' and 'polite' expression.

- Directive expression is a pragmatically sourced effect that conveys a wish. It is fairly decisive and straightforward, but not assertive (Verschueren 2003).
- Politeness has many forms. The one we suggest for computer speech, say over the telephone, is 'solidarity' politeness (Verschueren 2003). This is a type which tries to bring the listener into the situation.

The XML characterisation of expressive data structures is built around elements and their attributes. Here, our example for directive expression uses the element <directive_tone> with the phonological attributes *rhythm*, *AG* (accent group) and *intonation*. The attributes constitute variables subject to modification by procedures. With XML, notation procedures are themselves introduced using an attribute *procedure*, and in our example below each attribute has a set of procedures associated with it.

The relationship between the phonological and phonetic attributes is nonlinear, and this must be taken into account in the calculations on the phonetic dynamic plane. For example:

1 A relative value, say "low", on a phonological attribute will need conversion to a numerical value as the attribute 'moves' between planes, and which *particular* numerical value will vary with the environment.
2 A global instruction to render intonation as a particular contour is not appropriate for all sentences. If implemented repeatedly with variation, the machine voice would become monotonous and, after several sentences, listeners would report failure of naturalness.
3 For an angry emotive effect, assuming the correct intonation contour, rendering this contour by a global instruction to raise fundamental frequency overall by 10% may not be adequate in the same way.

Developments in Speech Synthesis   Mark Tatham and Katherine Morton
© 2005 John Wiley & Sons, Ltd.   ISBN: 0-470-85538-X

Consider also how the attributes of <directive_tone> or <polite_tone> appear on *certain* AG units but not across the *entire* IP which contains the string of AG units. Or certain rhythmic groups may be have their rate increased, such as those associated with the attribute *focus*, where the other rhythmic groups may in fact be slower. In addition there may be a *progressive* effect as the utterance proceeds, as with declination. A similar effect may be seen in speaking rate: rather than speed up or slow down an entire utterance, a more detailed and varying application of rules acting upon small units–and in conjunction with fundamental frequency (f0) changes as well–might well produce better results.

Looking at rhythm on the dynamic phonological plane for the sentence *Please hold the line*, spoken with cognitively sourced directive expression, we find (building on the previous diagram showing the relationship between the static and dynamic planes):

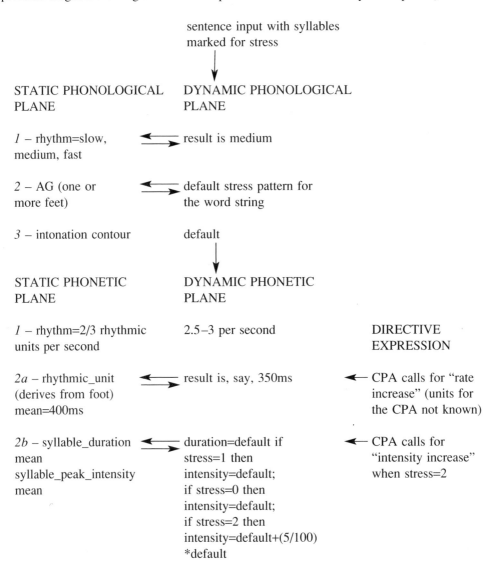

sentence input with syllables
marked for stress

STATIC PHONOLOGICAL   DYNAMIC PHONOLOGICAL
PLANE   PLANE

1 – rhythm=slow,   result is medium
medium, fast

2 – AG (one or   default stress pattern for
more feet)   the word string

3 – intonation contour   default

STATIC PHONETIC   DYNAMIC PHONETIC
PLANE   PLANE

1 – rhythm=2/3 rhythmic   2.5–3 per second   DIRECTIVE
units per second   EXPRESSION

2a – rhythmic_unit   result is, say, 350ms   ◄— CPA calls for "rate
(derives from foot)   increase" (units for
mean=400ms   the CPA not known)

2b – syllable_duration   duration=default if   ◄— CPA calls for
mean   stress=1 then   "intensity increase"
syllable_peak_intensity   intensity=default;   when stress=2
mean   if stress=0 then
                             intensity=default;
                             if stress=2 then
                             intensity=default+(5/100)
                             *default

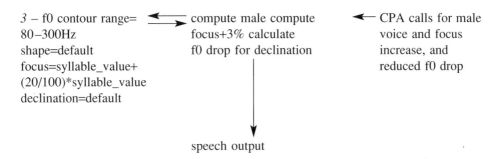

3 – f0 contour range= ⟵ compute male compute ⟵ CPA calls for male
80–300Hz   focus+3% calculate   voice and focus
shape=default   f0 drop for declination   increase, and
focus=syllable_value+   reduced f0 drop
(20/100)*syllable_value
declination=default

speech output

In the above diagram we find, in the rightmost column, CPA requirements for the utterance to hand. These feed to the dynamic plane (central column) which uses these requirements to modify basic specifications (including defaults) retrieved as needed from the static plane (leftmost column). In principle the CPA receives feedback from the dynamic phonetics, speech output (*via* listening), but is basing its requirements on information about dialogue context, and general utterance requirements. The values in the diagram are for illustration only.

31

XML Coding

The following is a sample coding of the diagrammatic information given in the previous chapter for directive expression.

- *Phonological static plane*–generalised code:

```
<default_tone
        rhythm="slow, medium, fast"
        AG=foot (+foot ...)
        stress_pattern="1(+0 ...)"
        intonation_contour="default">
```

The data structure on the phonological static plane characterises default tone, with three example attributes and their possible values. This is an abstract characterisation, and the ranges of values are there only to constrain selection procedures on the dynamic plane. Notice the expansion of AG to include the stress patterning within feet.

- *Phonological dynamic plane*–code for putative instantiation:

```
<default_tone
        procedure1="retrieve_rhythm_proc"
        procedure2="retrieve_stress_pattern_proc"
        procedure3="retrieve_intonation_contour_proc">
```

To compute the default tone for a particular utterance plan, the data structure is in effect a series of procedures which align with potential attributes to be located on the static plane together with their ranges of constraint. The section *procedure* conforms to the constraints and selects particular values from the ranges available. The result is an instantiated utterance plan constraint to have default prosody (within the example parameters given in our example–others are omitted for clarity). AG is omitted here since the computation involves the stress pattern within the feet contained within the AG.

- *Phonetic static plane*–generalised code:

```
<default_tone
        rhythm="default"
```

Developments in Speech Synthesis   Mark Tatham and Katherine Morton
© 2005 John Wiley & Sons, Ltd.   ISBN: 0-470-85538-X

```
rhythmic_unit=400
syllable_duration="mean"
syllable_peak_intensity="mean"
f0_contour_range=80-300
f0_contour_shape="rise_fall"
focus=syllable_peak_intensity+(20/100)*syllable_peak_
  intensity
declination="falling">
```

- *Phonetic dynamic plane*–code for putative instantiation:

```
<default_tone
    procedure1="compute_rhythm_proc"
    procedure2="compute_stress_pattern_proc"
    procedure3="compute_intonation_contour_proc">
```

31.1 Adding Detail

The computation required by many procedures will eventually be much more complex than illustrated here. This is because the relevant data is not yet available from experimental research in the field. Our purpose throughout the book is to illustrate an approach which we feel can be usefully developed along the lines sketched, but not to supply adequate data. These are sample procedures picked up from the data structures declared above.

The procedure retrieve_rhythm_proc–attribute *procedure1* on the *phonological* dynamic plane–might be

```
procedure retrieve_rhythm_proc
   {
      for syllable = 1 to j
         IF stress=1 THEN
              {
                    access static_plane
                    {
                          retrieve rhythm_range
                                   value="medium"
                          retrieve AG
                                   value="default"
                          retrieve stress_pattern
                                   value=1
                          retrieve intonation_contour
                                   value="default"
                    }
              } //characterises the words please and hold //
         IF stress=2 AND IF syllable="utterance_final" THEN
              {
                    access static_plane
                    {
```

```
                                        retrieve  rhythm_range
                                                  value="medium, ..., fast"
                                        retrieve  AG
                                                  value="default"
                                        retrieve  stress_pattern
                                                  value=2
                                        retrieve  intonation_contour
                                                  value="default"
                              }
                    }  // characterises the word line //
          IF  stress=0  THEN
                    {
                              access  static_plane
                              {
                                        retrieve  rhythm_range
                                                  value="low,  ...  medium"
                                        retrieve  AG
                                                  value="default"
                                        retrieve  stress_pattern
                                                  value=0
                                        retrieve  intonation_contour
                                                  value="default"
                              }
                    }  // characterises the word the //
          next  syllable
     }
  end;  // procedure retrieve_rhythm_proc //
```

A procedure 'compute_rhythm_proc' for directive speech on the *phonetic* dynamic plane might be

```
procedure  compute_rhythm_proc
     {
          for  syllable  =  1  to  j
                    IF  stress=1  THEN
                              {
                                        access  static_plane
                                        {
                                             retrieve  rhythm
                                             IF  CPA_input="1"  THEN
                                                     rhythm=rhythm+1
                                                     ELSE
                                                     rhythm=2.5
                                             retrieve  rhythmic_unit
                                             stress="default"
                                             IF  CPA_input="1"  THEN
                                                     intensity="default"
```

```
                        retrieve f0_contour
                        IF voice="male"
                                    range=(80 ... 200)
                        IF CPA_input="1" THEN
                          f0_drop=f0_drop - (15/100)*f0_drop
                  }
            }
        IF stress=2 THEN
            {
                  access static_plane
                  {
                        retrieve rhythm
                        IF CPA_input="1" THEN
                                rhythm=rhythm+2
                                ELSE
                                rhythm=2.5
                        retrieve rhythmic_unit
                        stress="default"
                        IF CPA_input="1" THEN
                        intensity=intensity + (5/100)*intensity
                        retrieve f0_contour
                        IF voice="male"
                                range=(80 ... 200)
                        IF CPA_input="1" AND
                        IF markup="focus" THEN
                                f0=f0+3
                  }
            }
        IF stress=0 THEN
            {
                  access static_plane
                  {
                        retrieve rhythm
                                rhythm=2.5
                        retrieve rhythmic_unit
                        stress="default"
                        retrieve f0_contour
                        IF voice="male"
                                range=(80 ... 200)
                  }
            }
      next syllable
    }
 end;  // procedure compute_rhythm_proc //
```

Notice that the code above could be collapsed further, but we have left it expanded for the
sake of transparency. The basic principle is that the procedures of the phonetic dynamic plane

assign values to abstract units derived on the phonological dynamic plane–and do so in the light of information supplied from the CPA. That the CPA is active is signalled in the code by 'CPA_input="1"', where "1" is Boolean. The units for intensity are decibels, and for fundamental frequency they are hertz. When domains wider than the syllable are retrieved, the values specified are the values for this syllable within the specified domain; thus, f0_contour has default or calculated values in Hz for each syllable within the contour's domain. We also use the value "default" to specify some norm.

We repeat: this is pseudo-code designed to illustrate the kind of thing that happens on the dynamic phonetic plane. It is not code which can be run or transported into an existing system. The reason for this is simple–we do not have the necessary data to feed into the code. Since the data is either scattered across the literature (and therefore derived in widely differing and potentially incompatible experiments) or not available at all, all we can do is suggest the kind of data that will be needed if speech synthesis is to have expressive rendering.

Bearing this point in mind, here is the code for the phonetic rendering of polite expression. The static and dynamic phonological planes as well as the static phonetic plane remain unchanged. The dynamic phonetic plane calculates new information dependent on instructions from the CPA to speak with a polite tone. What follows is the phonetic rendering:

```
procedure compute_rhythm_proc
    {
      for syllable = 1 to j
            IF stress=1 THEN
                {
                    access static_plane
                    {
                        retrieve rhythm
                        IF CPA_input="1" THEN
                                rhythm=rhythm - 0.5
                                ELSE
                                rhythm=2.5
                        retrieve rhythmic_unit
                        stress="default"
                        IF CPA_input="1" THEN
                                intensity="default"
                        retrieve f0_contour
                        IF voice="male"
                                range=(80 ... 200)
                            IF CPA_input="1" AND
                    IF (syllable=j - 1 or syllable = j) THEN
                            f0_drop=f0_drop - (15/100)*f0_drop
                    }
                }
            IF stress=2 THEN
                {
                    access static_plane
                    {
                        retrieve rhythm
```

```
                              IF CPA_input="1" THEN
                                      rhythm=rhythm - 0.5
                                      ELSE
                                      rhythm=2.5
                              retrieve rhythmic_unit
                              stress="default"
                              IF CPA_input="1" THEN
                      intensity=intensity + (2/100)*intensity
                              retrieve f0_contour
                              IF voice="male"
                                      range=(80 ... 200)
                              IF CPA_input="1" AND
                              IF markup="focus" THEN
                                      f0=f0 + 3
                      }
              }
          IF stress=0 THEN
              {
                      access static_plane
                      {
                              retrieve rhythm
                                      rhythm=2.5
                              retrieve rhythmic_unit
                              stress="default"
                              retrieve f0_contour
                              IF voice="male"
                                      range=(80 ... 200)
                      }
              }
      next syllable
  }
end;   // procedure assign_phonetic_rendering //
```

We assume that angry speech is biologically sourced and that directive or polite speech are cognitively sourced. For both types of source, we assume for the moment that both static and dynamic phonology are the same, that static phonetics is also unchanged, but that the calculations on the phonetic dynamic plane differ. The output settings in human speech change under biologically sourced expression, but operate within the ranges specified in the static phonetics. These limits therefore must be set to the widest possible for human speech.

Simulating the human expressive system using a computer model that is not articulatory based means devising a set of rules which will change the default values drawn from the static phonetics. Such a procedure is similar to characterising cognitively sourced effects, but differently motivated. In the first case, a set of dynamic phonetic rules operates within a range assuming all cognitive differentiations can be made within that range; in the second case, the baseline has shifted due to the *biological environment*. The parameters of the vocal tract now assume different *default* values. A complication may be that it seems to be possible to overlay cognitively sourced effects on biologically sourced events, as in 'polite anger', which may on the face of it seem absurd; but setting out the model in this way does enable biological and cognitive expression to be combined–this is an important feature. What is required is a set of constraints governing the set of possible cognitive overlays on the biological substrate; biopsychology will provide such constraints.

Following this principle, here is code for rendering biologically sourced angry speech:

```
procedure compute_rhythm_proc
    {
      for syllable = 1 to j
            IF stress=1 THEN
                {
                    access static_plane
                    {
                        retrieve rhythm
                        IF bio_anger="1" THEN
                                rhythm=rhythm + 1
                                ELSE
                                rhythm=2.5
                        retrieve rhythmic_unit
                        stress="default"
                        IF bio_anger="1" THEN
                                    intensity="default" + 0.5
                        retrieve f0_contour
                        IF voice="male"
                                range=(80 ... 200)
                            IF bio_anger="1" THEN
                    f0_contour=f0_contour + (10/100)*f0_contour
                            IF CPA_input="1" AND
                    IF (syllable=j - 1 or syllable = j) THEN
                            f0_drop=f0_drop + (5/100)*f0_drop
                    }
                }
            IF stress=2 THEN
                {
                    access static_plane
                    {
                        retrieve rhythm
                        IF bio_anger="1" THEN
                                rhythm=rhythm - 1
                                ELSE
                                rhythm=2.5
                        retrieve rhythmic_unit
                        stress="default"
                        IF bio_anger="1" THEN
                    intensity=intensity + (4/100)*intensity
```

```
                                        retrieve f0_contour
                                        IF voice="male"
                                                range=(80 ... 200)
                                        IF bio_anger="1" AND
                                        IF markup="focus" THEN
                                                f0=f0 + 5
                                }
                        }
                IF stress=0 THEN
                        {
                                access static_plane
                                {
                                        retrieve rhythm
                                                rhythm=2.5
                                        retrieve rhythmic_unit
                                        stress="default"
                                        retrieve f0_contour
                                        IF voice="male"
                                                range=(80 ... 200)
                                }
                        }
        next syllable
    }
end;  // procedure compute_rhythm_proc //
```

31.2 Timing and Fundamental Frequency Control on the Dynamic Plane

English is often described as a stress timed language (Lehiste 1970; Cruttenden 2001). This means that stressed syllables are perceived as spaced equally in perceived time, and unstressed syllables fit *within* this timing to form an overall pattern of rhythm. This impression of a regularly timed pattern is called *isochrony*, although it has not been possible to correlate this impression with measured acoustic data. The time duration between stressed syllables can be measured in millisecons, but it is not yet agreed that the concept of isochrony can be modelled adequately. We discuss this more fully in Part IX, but here we are concerned with how the concept might be programmed. Our theoretical position underlying practical synthesis is that there is indeed perceptual reality to the concept of isochrony and that this is what a speaker is aiming for, and therefore this is what a listener assigns to what is heard during the act of perception. There is a similar relationship seen in the concept of segment realisation–a speaker aims for a particular utterance plan and, whatever the variability in the actual soundwave, the listener reassigns this utterance plan to what he or she hears. This is what an act of perception *is*–the assignment of a hypothesis as to the speaker's intended plan.

The pseudo-code presented and discussed in what follows is for what we might call 'flat' or relatively expression-free speech. In practice this kind of output is undesirable and counter-theoretic. The reason for this is that it can never occur in real life–expression is all-pervasive and must be present to ensure the perception of naturalness. However, as we showed above, in practical situations, it is sometimes useful to compute expression-free speech internally, and then use it as the basis for computing speech with expression.

The following is an outline for a speech rhythm generator for synthesis, expressed in simple and general terms. This is far from being descriptive of an actual waveform. Remember that what we are really interested in, the proposed goal of speech synthesis, is to trigger or point to the best percept in the listener for matching up with the original speaker's intention. In our experience this will frequently mean that the synthesised soundwave is perceived correctly, but does not strictly match a soundwave produced by a human being.

31.3 The Underlying Markup

In phonological prosody we assume the existence of feet–units of potential rhythm characterised by equal length (abstract) and eventually equal duration (physical). Feet are defined as beginning with a stressed syllable. Hence an utterance can be segmented with respect to feet by introducing boundaries immediately before all stressed syllables. This is a single-tier linear process for the moment, though we feel strongly that the behaviour of feet in phonological prosody, and their physical correlates, rhythmic units, is better characterised as a hierarchical model.

Syllables are a hierarchical structure leading to a surface sequence of segments. Every syllable contains a vowel nucleus which, in English, can be preceded by zero to three consonants and can be followed by zero to four consonants. The phonological hierarchy of the syllable can be found in Chapter 1. Suppose a low-level concatenated waveform synthesiser and that we have assembled an appropriate string of units to match up with the extrinsic allophonic string characterising the utterance plan. It is likely to be the case that the reconstructed waveform will have inappropriate prosody. Our task now is to render the waveform in such a way that a listener perceives it to have an appropriate rhythm, and, later, an appropriate intonation.

We make some initial assumptions about the data, namely that the assembled waveform inherits, for rhythm, a markup with respect to

- syllable and segment boundaries
- stressing of syllables;

and additionally for intonation

- appropriate breaks (major and minor phrase boundaries).

This markup has been inherited from the segmental and wrapping prosodic plans generated in the phonological (high-level) part of the system. It has come to the newly assembled waveform *via* the selection process which has matched up the plans with the markup already existing in the system's database. It is important to minimise any mismatch here: the markup of the assembled excised 'real' units should optimally match the intended plans. As a practical point, we retain both markups for continuous comparison to identify areas of potential error.

In the SPRUCE system the syllable start boundary *for the purpose of computing rhythm* is marked from the time of the beginning of the first pitch period. This means that any waveform before this point falls within the previous rhythmic unit. The analogy is with

singing–stressed syllables in singing fall on the beat of the music. If the syllable begins with
a vowel, the beat is synchronised with the start of the vowel. If the syllable begins with a
so-called voiceless consonant, the synchronisation is still with the beginning of the vowel;
if it begins with a so-called voiced consonant, in English there is rarely actual vocal cord
vibration, and the beat is on the vowel, with the exception of nasals and liquids which
usually do have vocal cord vibration for much of their duration. Our system is perceptually
oriented; some researchers also taking the perceptual viewpoint synchronise on the 'perceptual
centre' of the syllable (Port *et al.* 1996). Both approaches share the idea that how syllables
begin is not relevant in rhythm–timing is tied to vocal cord vibration (onset for one group,
perceptual centre for the other). Figure 31.1 illustrates how we mark the start of a stressed
syllable, and therefore the start of a rhythmic unit–the physical counterpart of the
phonological foot.

Syllable marking indicates two levels, S and U–stressed and unstressed–as far as our phonological prosody is concerned. This corresponds to 'stress=1' and 'stress=0' in the code in
Chapter 30. Later we shall use F to indicate a stressed syllable which bears focus (sometimes
called *nuclear stress* or *sentence stress*) which corresponds to stress=2. Some phonologists
and phoneticians recognise other levels of stress, but this only complicates the issue here
and does not necessarily lead to an improved output, though it may have importance within
linguistics.

31.3.1 Syllables and Stress

We endpoint the start of a phonetic syllable with the beginning of the first period of vocal
cord vibration (as shown on the audio). The syllable then runs to the start of the next
syllable–assuming no pausing; that is, the default syllable 'environment' is stable and
idealised. The rhythmic unit–the physical correlate of the phonological foot–has its start
endpoint coinciding with the start of a stressed syllable. The rhythmic unit runs to the start
of the next stressed syllable–the default, under idealised conditions. In the pseudo-code below,
syllables stressed for focus (stress=2, see above) are treated as normally stressed for the sake
of clarity.

Here is the pseudo-code for locating the start of a rhythmic unit based on the phonological syllable marking inherited by the reassembled units from their location in the marked
database:

```
procedure locate_rhythmic_unit_start
begin
      for syllable = 1 to j
              IF stress = 1 THEN
                    begin
                             proc_locate_first_period;
            // returns t_period, the time of the first period from the utterance start //
                             marker := t_period;
                      // sets boundary marker //
                             end;
        next syllable
end; // procedure locate_rhythmic_unit_start //
```

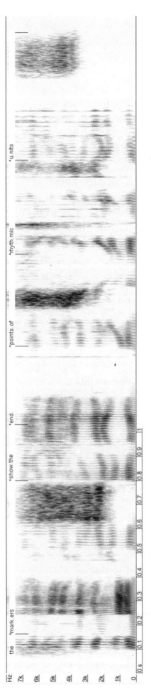

Figure 31.1 Spectrogram of the sentence *The markers show the endpoints of rhythm units*. Within-word syllable boundaries are indicated by a point (.) in the orthographic labelling. Vertical marks indicate the boundaries of the various rhythmic units in the utterance as indicated by the start of each stressed syllable. Stressed syllables have an asterisk.

Suitably enhanced, this code will collect a string of start times for rhythmic units and mark the assembled waveform. At the same time a suitable procedure will similarly mark the start of each syllable–each begins on the first identifiable period (anything before goes to the previous syllable).

31.3.2 Durations

The idealised default condition assumes all rhythmic units to be of equal duration. This is justified by the perceived equal length of phonological feet. The fact that they are never found to be of actual equal duration in real speech is catered for by allowing the default values to be deformed either under control of the CPA (which has its principled reasons for changing the values) or by randomly generated distortion. Apportioning duration to syllables within the rhythmic unit is rule-governed in this system:

1 If there is one syllable (by definition stressed) it shall occupy the entire time.
2 If the stressed syllable is followed by one unstressed syllable, it shall occupy 2/3 of the time; the remainder is taken up by the unstressed syllable.
3 If the stressed syllable is followed by two unstressed syllables, it shall occupy 1/2 of the time; the remainder is taken up equally by the two unstressed syllables.
4 If the stressed syllable is followed by three unstressed syllables, it shall occupy 1/3 of the time; the remainder taken up equally by the three unstressed syllables.

Reminder: the values are exemplar defaults and are the starting point for making the system work. They are tweaked by CPA intervention in an actual system. As remarked earlier, the data for this is unavailable, revealing, along with how rhythmic unit timing is distorted, important research questions.

Here is some pseudo-code for determining the duration of individual phonetic syllables within each rhythmic unit:

```
procedure compute_syllable_durations
begin
        counter := 0;
        // set counter to find stressed syllables //
        for rhythmic_unit = 1 to j
            for syllable = 1 to i
                    IF (stress = 1 OR stress = 2) THEN
                    begin
                            counter := counter + 1
                            IF counter > 1 THEN exit;
                            // the rhythmic unit is already ended //
                            number_of_syllables := 1
                    end;
                    IF stress = 0 THEN number_of_syllables :=
                                    number_of_syllables + 1
```

```
                     next syllable
                     for syllable = 1 to i
                             IF number_of_syllables = 1 THEN
                                     syllable_duration := unit_duration
                             IF number_of_syllables = 2 THEN
                             begin
        IF stress = 1 THEN duration := (67/100)*unit_duration
        IF stress = 0 THEN duration := (33/100)*unit_duration
                             end
                             IF number_of_syllables = 3 THEN
                             begin
        IF stress = 1 THEN duration := (50/100)*unit_duration
        IF stress = 0 THEN duration := (25/100)*unit_duration
                             end
                             IF number_of_syllables = 4 THEN
                             begin
        IF stress = 1 THEN duration := (34/100)*unit_duration
        IF stress = 0 THEN duration := (22/100)*unit_duration
                             end
                   next syllable
           next rhythmic_unit
   end
```

In assigning duration to syllables we are in fact allocating time to the syllable and, based on its underlying structure, this time is apportioned to the segments within the syllable. Notice that we are not building up the time for a rhythmic unit based on a linear accumulation of default times for individual segments. The linear bottom-up approach has never provided really satisfactory results and is counter-theoretical. The structure is always hierarchical and captured thus:

```
<rhythmic_unit>
        <syllable stress = 1>
                <component_segments/>
        </syllable>
<syllable stress=0 range="0-4"/>
//assumes 0 to 4 unstressed syllables//
                <component_segments/>
        </syllable>
</rhythmic_unit>
```

31.4 Intrinsic Durations

Nevertheless, syllables are often described as having an 'intrinsic length' determined by summing the actually occurring durations of the component allophonic segments. The problem is that these durations will differ for the same allophone in different segmental and prosodic contexts. The syllable [bɪt] (with a short vowel) has a shorter intrinsic duration than the syllable [bɑt] (with a long vowel). One suggestion has been that syllable durations will vary less than what we observe here for isolated syllables because the need to maintain rhythm overrides the intrinsic duration measured in isolated words. There is scope for increasing duration; for example, the voice onset time in [bɪt] is flexible–a longer voice onset time is unlikely to interfere with perception in an utterance since context contributes to inter- pretation. This idea needs much further investigation; what we are saying is that perceptual constraints can signal whether durational rendering can be varied sometimes to accommo- date production constraints. The situation is very complex, and this is one reason why we have found it necessary to introduce the idea of supervised rendering.

In conversation, rhythm based on syllable strings is often interrupted by inserting a pause.

- Does the pause highlight the following stressed syllable?
- Alternatively, is it perhaps a silence equivalent to an unstressed syllable and functions to maintain rhythm?
- Is it inserted at the end of the previous accent group, or between two accent groups or rhythmic units, perhaps maintaining the optimum duration of the accent groups in the conversation?

These are questions which must be addressed when building a rhythm generator for synthesis; there are knock-on effects with intonation and focus.

31.5 Rendering Intonation as a Fundamental Frequency Contour

The input to the intonation rendering algorithm we devised for SPRUCE requires a hierar- chical, XML-style markup consisting of the labels S, U, F (stressed, unstressed, focus) on syllables, and required inclination on phrases and sentences, L–H, or declination, H–L. To accommodate the markup we have also to establish phrase/clause boundaries, sentence end- points, the type of sentence–WH-question, declarative, question–any cognitive expressive contribution, any biological sourced expression/emotive contribution, and any contribution from the CPA. The assignment of the markup and the labels themselves are discussed in Chapter 4.

1 Determine the available fundamental frequency range for this utterance (data comes from the static phonetic plane).
2 Find the focal nuclear syllable (marked F).
3 Assign to F the highest available frequency within the permitted range for this utterance. This results in inclination from the sentence start to F, and declination from this point onward.

Expressive effects can be added by changing the values on 'some_value'. For example, the inclination, focus and declination values can be modified to systematically take into account the need for polite or even angry speech. The illustrations are given using the basic JSRU text-to-speech (Holmes *et al.* 1964) implementation for production of the so-called *neutral* speech that the unmodified JSRU system produces.

The following is the algorithm for moving between the prosodic utterance plan, and the segmental utterance plan fitted to it, and the physical rendering of a typical sentence. This is a scaled-down implementation from the SPRUCE version of the Tatham and Morton prosody model (see Chapter 32). In what follows, S means <syllable stress=1>, U means <syllable stress=0>, and F means <syllable stress=2>. A WH-question is a question beginning with a word like *who*, *what* or *how*. Numerical values assigned to the fundamental frequency are in Hz and are averages for a single typical sentence–any particular instance will have different values. Any expressive modifications will also alter these values.

1: Assign *Basic* f0 Values to All S and F Syllables in the Sentence: the Assigned Value is for the Entire Syllable

a. IF the first syllable marked S is preceded by more than one unstressed syllable THEN
f0 := 110 ELSE f0 := 105.

b. IF this is the last syllable marked S THEN
f0 := 85.

c. For each syllable marked S (except the last one)
f0 := previous_f0 - drop;
where drop = round [(f0 during first S - f0 during last S) / $(n - 1)$],
where n is the number of syllables marked S in the sentence.

d. f0 during syllable marked F is 7 above the value assigned by 1, 2 or 3 above.

2: Assign f0 for All U Syllables; Adjust *Basic* Values

a. The f0 value for all U syllables is 7 below the f0 value for the next syllable (whether S or U) if there is only one U syllable between S syllables (or between the start of the sentence and the first S syllable) ELSE the value is 4 below.

b. IF the sentence is paragraph_final=1 (true) THEN
for the last S syllable or F syllable, decrement by 4 every 80ms (i.e. the first 80ms have the value assigned in 1 above, the next 80ms have this value less 4, and so on).

c. IF the sentence is paragraph_final=0 (not true) THEN
the f0 for the last S or F has the first 20% of its vowel duration set to 77 and the remainder to 84, irrespective of values assigned above.

d. IF the sentence is paragraph_final=1 (true) THEN
any U syllable after the last S or F syllable has f0 set to 4 less than the final value during the S syllable: this value to be held constant during the entire syllable;
ELSE

IF the sentence is paragraph_final=0 (not true) THEN
any U syllable after the last S syllable or F syllable has f0
set to 77.

3: Remove Monotony

a. Except for the final S syllable and any U syllables following
it, decrement f0 by 4, 50% of the time through every syllable
(S or U).

4: For Sentences with RESET, where a RESET Point is a Clause or Phrase Boundary

a. For the S or F syllable preceding a RESET point, lengthen the
vowel part *only* of the syllable by 50%.
b. The first 25% of the vowel part of this S syllables should be
left at the f0 value assigned by earlier rules, but the remainder
of the vowel to the end of the syllable should be dropped by
7 for any values previously assigned.
c. IF RESET is followed by a conjunction or function word (determined by PARSE), THEN decrement f0 by 4 throughout *this* word
(as opposed to syllable): this is the word preceding the RESET
point.
d. f0 values for any S or F syllable from a RESET to the end of
the sentence need to be recalculated:

> **i.**   last S syllable f0 stays at 85 if S, or 92 if F;
> **ii.** increment f0 during the first S or F after the RESET by 7;
> **iii.** adjust S syllables between this first S after the RESET
>   and the last S to decrement smoothly: that is, by calcu-
>   lating a new value for the continuous decrement under 2d
>   above; but, assume the last S to be 85, even if has been
>   raised to 92 because it is F;
> **iv.** modify f0 for all syllables between the S syllables as in
>   2a above.

32

Prosody: General

Prosody is often defined on two different levels:

- an abstract, phonological level (phrase, accent and tone structure)
- a physical phonetic level (fundamental frequency, intensity or amplitude, and duration).

Using the approach of coarticulation theory (that speech consists of conjoined segments which overlap in some way), we can observe that

- the durations of segments vary
- there are runs of these segments separated by pauses.

These temporal phenomena are the basis of prosody. If segments are classified by labelling them symbolically according to their underlying phonemes or extrinsic allophones, we can see that they vary in duration systematically. So, for example, in English among the vowels the sound [ɑ] is almost always greater in duration than [æ], [u] has a greater duration than [ʊ], [i] greater than [ɪ]. These differences are systematic and contribute very much to our understanding of the rhythmic structure of utterances. The prosodic structure of the language provides the framework for segmental renditions. To enable comparison between different approaches to the relationship between prosody and segmental structure, we shall use this characterisation as our starting point; it does, though, fall a little short of our final view, that prosody forms a tight wrapper. One way in which we differ is that our approach enables prosody itself to be wrapped by expressive content–a concept which is essentially 'tacked on' to utterances in other models. It is enough to say for the moment that prosody forms a tight system in its own right and is not dependent on the segmental makeup of utterances. What we leave open for the moment is the exact dominance relationship between the two.

It is generally felt that from the listener's point of view one purpose of prosody is to cue syntactic boundaries and to indicate something of underlying syntactic structure. Thus major syntactic groupings are marked by systematic patterning of the prosody–for example, the slowing down of the repetition rate of rhythm units (perceived as a distortion of isochrony) toward the end of a sentence, and the accompanying general fall in fundamental frequency (perceived as a lowering of the intonation contour). Slowing rhythm and falling intonation signal an imminent sentence end, especially when they come together. In addition we find that prosody is often used to disambiguate sentences, by introducing some kind of stress, for example, on particular words or groups of words. Similarly when the pragmatic context is

Developments in Speech Synthesis   Mark Tatham and Katherine Morton
© 2005 John Wiley & Sons, Ltd.   ISBN: 0-470-85538-X

ambiguous, stressing and rhythmic changes help the listener choose from what might have been an unmanageable set of hypotheses as to the *intended* pragmatic context.

It would follow that, from the speaker's point of view, part of the function of prosody is to signal syntactic structure. It is for this reason that most researchers in text-to-speech synthesis include a syntactic parse of the incoming text as a preliminary to the assignment of phonological prosody. We shall see that phrase breaks and the like figure prominently in this approach.

32.1 The Analysis of Prosody

The analysis of prosody is important (Monaghan 2002) in speech synthesis because it gives us the basis for marking prosodic effects around our utterance plans (phonological prosodic processing) and later to arrive at suitable rendering strategies for the marked prosody (phonetic prosodic processing). There are two approaches in the prosody literature:

1 Create an abstract descriptive system which characterises observations of the behaviour of the parameters of prosody within the acoustic signal (fundamental frequency movement, intensity changes and durational movement), and *promote the system to a symbolic phonological role.*
2 Create a symbolic phonological system which can be used to input to processes which eventually result in an acoustic signal judged by listeners to have a proper prosody.

We must be careful to avoid circularity here, but essentially the difference between the two is the extreme data orientation of the first approach and the very abstract orientation of the second. They relate, of course, *via* the acoustic signal–in type 1 the source of the model is the acoustic signal, and in type 2 the outcome of the model is a set of hypotheses about the acoustic signal. Both approaches require a representational system or the means of transcribing prosody. Type 1 transcribes the prosody of *existing* acoustic signals, but type 2 transcribes a prosody for *predicting* acoustic signals. The prosody of *classical phonetics* falls into type 1, and the prosody of more recent systems such as ToBI (see below) falls, at least in principle, into type 2.

One way of looking at how the approaches differ is to consider whether apparent differences between data objects are significant; and in this context it means asking whether the difference between two fundamental frequency curves, for example, plays a linguistic role–perhaps contrasting for the listener two subtle, but systematically recognisable, shades of meaning. This is precisely the problem phonology and phonetics have tacked from the segmental perspective. The consensus here is that some kind of highly abstract unit should be recognised as deriving less abstract more surface variants–this unit is, of course, unpronounceable and has primarily, in the earliest but one of the clearest versions of transformational phonology, a contrastive function designed to give a unique representation for each morpheme. Such a representation cannot itself be found on the surface, and its 'existence' is given only by the existence of its derivatives. This approach leads to the phoneme/ allophone hierarchy in which we are able to discriminate between two types of allophone, *extrinsic* and *intrinsic*–those which are phonologically derived and those which are phonetically derived respectively. In the context of this book, remember the extrinsic allophonic representation is used to characterise the utterance plan, which is the final surface representation of the utterance prior to phonetic rendering or passing to a low level synthesiser.

Prosody parallels the segmental approach. The ToBI markup, for example, in effect provides a characterisation of intonation equivalent to segmental extrinsic allophones. One of the problems with this markup for text is how to provide a set of rendering processes through an intrinsic allophonic level as far as the actual physical exponents. On the other side of the coin it is equally difficult to provide an explicit set of procedures for using the system to mark up the audio–hence the inadequacies of automatic methods for doing this. None of the systems we have at the moment is quite explicit with respect to how it relates to the acoustic signal in either direction, analysis or synthesis.

In principle, highly abstract markup of the ToBI kind would be satisfactory provided it could be transparently related to the acoustic signal. The trouble is that we know even less about how to render high-level prosodic representations than we know about rendering something as abstract as a phoneme from the segmental perspective. There are recognisable levels of allophone in prosody (i.e. identifiable variations), just as there are for segments. And we can hypothesise different types of variability, just as with segments. Automatic conversion from markup to soundwave is hampered by the need to progressively expand the representation including more and more variability, although we do not understand as yet the types or sources of all the variability.

In recent times most of the discussion has been about transcription of prosody, and this has been equated with its analysis. It has been considered important to reflect the analysis as accurately as possible in the transcription system. The risk for us working in synthesis is whether the analysis is appropriate for capturing all that is needed for re-rendering, and the question is whether the original rendering can be reconstructed from the analysis as transcribed.

It may be necessary to rethink the linguistics supporting the analysis. Linguistic analysis is a data reduction process–information is removed in layers according to its salience. Thus minor, apparently random, acoustic variations are quickly removed; variations constrained by the design of the vocal tract and its aerodynamics (coarticulatory effects) are removed in the next layer; language-specific but semantically irrelevant variation is removed next–and we are left with the underlying 'essence' of the utterance, its phonemic (or similar) representation. The difficulty is that *re*-synthesis may not be a mirror image of the analysis process.

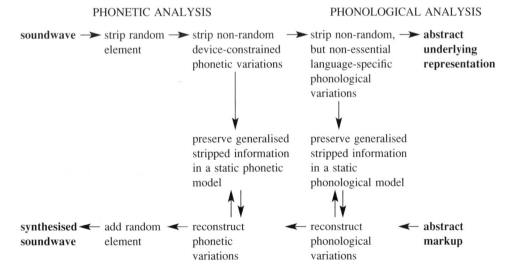

The diagram illustrates how linguists take a soundwave and construct firstly a phonetic model by removing variability down to coarticulatory effects; then secondly construct a phonological model by removing language-specific variability; and move finally to a completely abstract minimal representation of what is said to underlie the utterance. Current thinking in synthesis, as shown by database markup in concatenated waveform unit selection systems, is to take these underlying representations, or a representation at a previous level (or, worse, an unprincipled mixture) and attempt to reconstruct the soundwave by successively consulting phonological and phonetic models.

32.2 The Principles of Some Current Models of Intonation Used in Synthesis

32.2.1 The Hirst and Di Cristo Model (Including INTSINT)

The Hirst and Di Cristo model (Hirst and Di Cristo 1984; Di Cristo and Hirst 2002) incorporates four levels of representation or four components.

1 The *physical level* characterises universal constraints. These are the random elements, and perhaps some phonetic elements in the diagram above.
2 The *phonetic level* characterises phonetic constraints and is designed to act as a link between cognitively sourced phonological properties and physically sourced phonetic properties. It works in terms of a string of target values along the fundamental frequency curves.
3 The *surface phonological level* shows phonological derivations and how they are determined. It 'captures' the target values identified at the phonetic level in a symbolic transcription.
4 The *deep phonological level* is highly abstract and symbolic, characterising what underlies the surface phonological level.

Firstly it is not entirely clear whether they mean *level* or *component*. A level might be a 'level of representation', or the representation itself, whereas a component would be responsible for deriving a level from some other level.

The *physical level* will include those properties of acoustics, aerodynamics, motor control etc. which constrain the possibilities for the use of these systems in language–they are for the most part uncontrollable universals. To be effective the model needs to identify the intrinsic ranges of variability of the various elements at these levels. Within our own philosophy it would also be necessary to identify and characterise those universal effects which attract and succumb to cognitive intervention–not only which, but how much and with what repeatability–since cognitive intervention in physically dominated universals is a pillar of our suggested model.

The *phonetic level* stands between anything cognitive or phonological and the physical or universal system. This is intended to be a kind of abstract or idealised symbolic representation of the physical side of things rather than the cognitive side of things–but couched in such a way that it can look in both directions explicitly to provide a useful interface between the two.

The *surface phonological level* corresponds in our model to the level at which the plan appears. It is characterising cognitively derived instantiations. The use of their term prompts the observation that this is probably not a component.

The *deep phonological level* is about the schema to which any dynamic prosody must conform–the knowledge of prosody in a human being, or the characterisation of all of prosody in the model. This is the level for abstract markup.

32.2.2 Taylor's Tilt Model

Like most of the leading intonation models, the *Tilt* model (Taylor 2000)–which is derived from Taylor's *RFC* (rise–fall connection) model (Taylor 1995)–calls for tiered representations obeying the usual rules of constraint and data reduction (removal of redundancy) to achieve a model intended to account for perceptually distinct utterances–a principle adapted from analytical linguistics. Note that these are *perceptually* distinct rather than *linguistically* distinct. The approach allows for the possibility that there may be perceptually important intonational phenomena which are not linguistically significant, and it may be necessary to include these. Linguistic relevance, however, remains a focus for this model.

A distinguishing feature of the RFC and Tilt models is a robustness which derives from their computational approach. That the model will compute is at worst a confirmation of its internal coherence and at best enables explicit logical and procedural relationships between elements and data structures within the model. This leads, computationally, to an easier relationship between abstract symbolic markup and physical instantiation, a property which is important in models underlying speech technology. The Tilt model was developed using a bottom-up approach–that is, beginning with the acoustic signal and proceeding through to the more abstract symbolic representation.

The Tilt model seeks to *extract* from the acoustic signal (or, better, in our approach *assign* to the signal) a linguistic significance, but at the same time be able to generate the acoustic signal back from what is linguistic, as shown in the generalised diagram above. To do this requires more information than is contained within any linguistic representation–it needs information about the physical system and how it might actually be used. Taylor recognises that the way in which prosody is used in utterances involves rather more than the coverage afforded by traditional linguistics. We assume that Taylor sees the need to extend traditional descriptions to include somehow a characterisation of expressive and other effects.

32.2.3 The ToBI (Tones and Break Indices) Model

ToBI (Silverman *et al.* 1992) is a model which provides for a highly abstract markup of text or waveforms to indicate linguistically relevant features of prosody. The model is fairly explicit, and rests on a number of criteria.

1 The model is to be a characterisation of phonological intonation, rather than its physical correlates.
2 Consequently it should be a symbolic representation of what underlies the acoustic signal, not a symbolic representation of the signal itself.
3 What the model characterises should be in the spirit of linguistics. It should represent distinctive information of a symbolic nature rather than any automatically detectable physical parameters.

What is important here is the emphasis on a highly abstract characterisation of intonation, and the explicit rejection of the physical signal as the object of description. This does *not* mean that analysis cannot take place with the signal in mind; what it does mean is that the analysis is *not* of the signal itself.

A consequence of the explicit strategy for ToBI is that its characterisation of the intonation of an utterance has then to be properly rendered as the specification for the associated acoustics. There is no reason for not using a ToBI representation in a synthesis system provided it is realised that it

- is abstract, and high-level
- is the result of analysis
- may need reformulating in a synthesis system
- requires a low-level rendering.

However, because of the distinctiveness constraint it is not guaranteed that sufficient detail can be marked to avoid reinterpreting the markup before normal phonetic rendering processes are applied. Another way of saying this is that there is the possibility that not all phonological information is recorded in the markup. Careful application can probably avoid this problem.

32.2.4 The Basis of Intonation Modelling

In our references to prosody we repeatedly make a distinction between phonological and phonetic prosody, and it is no surprise that researchers are divided as to where in this hierarchy intonation models fit. In prosody, as in segmental modelling, a phonetic characterisation will emphasise the dynamic aspect of the signal–especially its continuous-ness; whereas a phonological characterisation, because it is always symbolic, favours discreteness in the representation. The static/dynamic dichotomy remains; either treatment, phonetic or phonological, can be static or dynamic along the lines explained in Part VII. One or two researchers summarise the phonological/phonetic distinction in terms of the quantitative emphasis of phonetics and the qualitative emphasis of phonology. Carmichael (2003) has a comprehensive statement and evaluation of the two approaches.

- *Phonetic models* of intonation provide an explanation of the intonational features of the acoustic signal, especially the fundamental frequency. It will not be expected to include random variability, but should include all systematic variability introduced through phys-ical constraints. However, because it is based on physical features it cannot be expected that an explanation of linguistic features should be included. At best, the model explains the acoustic signal given an appropriate linguistic characterisation as its input–the prosodic wrapper of the phonological plan, in our terminology. In principle it should be possible to reconstruct the acoustic signal (but without non-systematic variability) from the model's output. Note that the output cannot be tested by a listener–listeners perceive by bringing additional information to the signal and assigning a symbolic representation. Because of this additional information, over which the phonetic model has little control, we cannot evaluate perceptually the accuracy of any acoustic prosody specified by the model. Perception is particularly adept at repairing damaged signals; so, unless we know the extent of the repair, we cannot estimate the damage and consequently cannot estimate the signal. Evaluation of the output of a phonetic intonation model within a synthesis system probably has to be by inspection of the signal and an objective comparison between it and a similar human signal.

- *Phonological models* may consult the acoustic signal, but only to prompt an immediate data reduction process for deriving a discrete symbolic representation. Because of its phonological nature, such a model is of direct relevance to listeners and perception. What listeners assign to an acoustic signal is a phonological representation–a phonological model is often about the processes involved in this assignment. This makes phonological prosodic models relevant to both synthesis and automatic speech recognition. In synthesis the focus is on prosodic interpretation of semantic, syntactic and pragmatic features to specify the prosodic wrapper for the utterance plan. In automatic speech recognition the focus is on the processes involved in assigning an appropriate symbolic representation to a waveform such that semantic, syntactic and pragmatic features can be processed.

The emphasis with phonological models is therefore on what is linguistically important–the underlying representation–which corresponds also to what is perceptually relevant; they de-emphasise the actual signal itself and work only on symbolic representations. Phonetic models emphasise the acoustic signal, explaining the movement of fundamental frequency through the utterance–but not, in a truly phonetic model, in terms of the linguistics of the utterance; they de-emphasise the linguistic features which may underlie the signal. Both phonological and phonetic models of intonation can be linearly oriented or hierarchically oriented, linear models emphasising the surface context of events and hierarchical models emphasising the underlying structure and the contexts revealed within it. It seems to us that most phonological and phonetic events are explained only in terms of an underlying structure which is more complex than any purely linear analysis could afford.

32.2.5 Details of the ToBI Model

ToBI is a high-level or phonological prosodic model stemming from principles of autosegmental phonology (Goldsmith 1976). In this approach an intonation contour specifies a strictly linear concatenation of so-called pitch events. Thus ToBI attempts to explain intonational data structures from a surface perspective. Occasionally there is reference to levels, but formally the model is linear. In this section we comment on some of our understanding of ToBI properties as intended, and also as assumed by users–the two do not always match. Our intention is not to characterise the detail of ToBI (see Silverman *et al.* 1992).

1 Global intonational contours which span stretches of speech are explained in terms of the pattern built up by local events or local patterns of a string of events. The 'continuous' global event–the utterance intonation contour–is, in effect, the outcome of interpolating local events. The word *continuous* is our way of expressing the internal integrity of the string of symbols assembled during a ToBI analysis. The internal integrity is captured by the explicit surface patterning within the string, without recourse to any hierarchically organised underlying data structure.

2 Local pitch events, such as pitch accent tones and boundary tones, are regarded as 'targets' along the global contour. Compare this with the way SSML marks up a global intonation

contour (see Part V) as a movement of fundamental frequency interpolated between identified points along the way.

3 Local pitch events, characterised by ToBI high-level symbolic representation, are nevertheless able to be linked with actual physical events. But it is theoretically impossible to link certain aspects of a time-governed physical signal with an aoristic static characterisation. Loosely delimited stretches of signal can at best be associated with ToBI's abstract symbolic characterisation. This is not a criticism of ToBI which is truly in the spirit of phonological prosody; it is, however, a general warning against trying to identify temporal events in abstract representations derived in the spirit of linguistics.

4 Although considered an analysis tool (i.e. the symbolic representations are derived from the actual physical events), ToBI markup is often considered by synthesis researchers to be adequate for representing projected utterance fundamental frequency contours. Applied this way the most that could be expected is a smoothed or idealised exemplar physical contour, but not an adequate example of a contour which might really have been produced by a human being. The theoretical assertion rests on the fact that ToBI involves a lossy reductionist approach which denies the possibility of accurate reconstruction. There is a strong parallel here with attempts to produce utterance plans wholly from a static phonology; at best all that can be produced is an idealised exemplar plan bearing only a symbolic relationship to what actually does happen in the physical world.

As a model of intonation, ToBI is useful in high-level synthesis as a relatively explicit characterisation of *phonological* intonation. How this relates to phonetic prosody is not the business of the model *per se*, although ToBI markup of a soundwave is based on some rules, though probably not explicit enough to make for reliable and unambiguous symbolic characterisations as might be required in automatic speech recognition–the system is designed for use by human transcribers of an existing soundwave. Because of the lossy nature of symbolic representations of this nature, recovery of fundamental frequency contours from a ToBI markup by a low-level synthesiser is at best dubious, though it might constitute an unintended and therefore unfair use of the system. At worst ToBI does rely to a certain extent on the transcriber's tacit knowledge of the relationship between soundwave and cognitive representation, making this a hybrid system–part objective and part subjective.

A derivative of ToBI, Intonational Variation in English (IViE) (Grabe *et al.* 2000), purports to add to its parent system to enable greater scope for marking variation in intonation. It was designed to assist differentiating between phonological and phonetic variation, since, it is claimed, some variations are phonological in one accent of English but phonetic in another. However, the model is still linear, thereby missing any insights which might come from considering the underlying structure of prosody, and still rather impressionistic in terms of how it is applied to a soundwave. Variability in speech, whether it be segmental or prosodic, has multiple sources–phonological, cognitive phonetic and phonetic. These sources derive variability in the 'basic' pronunciation, in expressive or emotive speech, in dialectal or accent pronunciation, in idiolectal versions–it is highly likely that a tiered model will prove of greater value than a flat model which can really simply list alternatives in some instances. The fact that some underlying phonological features in expression may interact with more surface cognitive phonetic accent variation, for example, really cannot be captured in a linear characterisation of prosody.

Any model which relies on transcriber judgement for analysis, even if it is only to a small extent, will have some problems with objectivity and inter-transcriber comparability. But this becomes a serious matter when, as with IViE, there is a deliberate attempt to build a model suitable for inter-accent or inter-dialect variability. It is unlikely that the transcriber will be a true native speaker of more than one accent–though the person may think that he or she is good at imitating more than one. For this reason the person's judgements fall short of ideal, and there is bound to be colouration of his or her assignment of symbolic representations. It is has been shown many times that no matter how well trained or practised the transcribers, fully objective assignment of symbolic representations to waveforms is doubtful even for a native speaker, and it is surely doubly doubtful for non-native speakers. True bi-dialectal ability is probably as rare as true bi-lingual ability. There is even scope for the view that training actually reduces objectivity by training-in implicit data reduction techniques.

32.2.6 The INTSINT (International Transcription System for Intonation) Model

INTSINT (Campione *et al.* 2000; Di Cristo and Hirst 2002) is a linear phonetic intonation model designed for marking intonation events on a waveform. The basis for the markup comes from direct measurements of the changing fundamental frequency of the signal. Designed for use across different languages, the system does not call for any prior knowledge of how the phonological prosody of the language works. It is perhaps best to regard INTSINT as a tool for gathering phonetic phonological data, rather than as a true model of intonation; true intonation models cannot really escape having a true phonological component.

In common with SSLM, INTSINT regards a fundamental frequency contour as the result of interpolation between isolable pitch targets. In this sense the INTSINT model is directly related to the proposals of 't Hart and Collier (1975). It is not clear whether an analogy with coarticulation theory can be developed here. Coarticulation theory applies to segmental rendering and postulates time- and inertia-governed movement between the targets of successive segments, allowing for the concept of 'missed targets', the curve exhibiting under- or overshoot. If this concept were applied to phonetic intonation, or indeed to prosody in general, a model could be developed in terms of phonetic prosodic targets which would not only become linked in a general movement of the parameters between them, but also be subject to coarticulation-style under- and overshoot. The INTSINT developers do not claim this explicitly; but, as with coarticulation theory for segmental representations, it would form a compatible basis for dealing with prosodic features and variability in the acoustic signal.

INTSINT is important as a model because it can be related directly to the acoustic signal. The system is for analysis rather than synthesis in the first instance, though clearly it is in the right area of the chain of events in either direction (toward or away from the waveform–production or perception, synthesis or recognition)–to have the potential for capturing the kind of variability displayed in the signal which is not normally represented in the symbolic representations of phonological prosody models. The reason for this is that if symbolic representation is necessarily lossy, at least there is the possibility of establishing a set of meta-constraints as a by-product of the reduction process which can be held ready for a truer reconstruction of the soundwave than might otherwise be possible.

32.2.7 The Tatham and Morton Intonation Model

The linguistically oriented model of prosody developed by Tatham and Morton (Morton *et al.* 1999; Tatham *et al.* 2000) is incorporated in the SPRUCE high-level synthesis system, and shares with Taylor's RFC and Tilt models an insistence on full computational adequacy (Tatham and Morton 2004). The model incorporates a phonological intonation model which is nonlinear in nature, paying special attention to the dependencies between symbolic representations and their physical renderings. One reason for this is that the model implements the principles of cognitive phonetics with respect to cognitive control of intrinsic phonetic or rendering phenomena, whether they are segmental or prosodic in origin. Another way of saying this is that the model recognises that there are some phonetic processes (distinct from phonological processes) which have true linguistic significance.

The model makes the claim, and is designed to capture it, that many of the physical processes involved in speech production are cognitively represented. For this reason the approach can be said to move forward research ideas in synthesis by insisting on tying in with both the general acoustics and aerodynamics of an utterance. It does not simply attempt a direct link between the phonological representation and the fundamental frequency contour, as did Pierrehumbert, for example (Pierrehumbert 1981). The general philosophy derives from cognitive phonetics and is summarised in Tatham *et al.* (1998, p. 1):

> Our underlying principle is that in the human being there are physical processes intrinsic to the overall speech mechanism, and that some of these processes are open to cognitive representation–such that they are able to enter the domain of language processing.

Because the T&M model claims that the relationship between production and perception is one involving knowledge of *both* components within *each* component, there is a clear path toward explicitly relating variation in the soundwave to the cognitive representations and processes involved in both production and perception. The model requires that a linguistically relevant acoustic event (including those variations which are linguistically relevant) be explicitly identified in both production *and* perception. Furthermore any two linguistically distinct phenomena (prosodic or segmental) have to be reliably and consistently produced and perceived as equally distinct. There is no necessity for a linear correspondence between physical processes, and once their cognitive representation is established it becomes part of language processing. The overall claim is that with strict adherence to the correspondence between cognitive and physical phenomena we end up with a phonological model, but one which consistently and transparently explains phonetic outcomes. We claim this is a fundamental requirement for prosody models in speech synthesis.

Units in T&M Intonation

In the T&M model of intonation, scaled to apply in SPRUCE high-level synthesis, we mark lexical stress, sentence and phrase boundaries, and focus. Following many other researchers, the assignment of intonation is dependent on the syntactic structures of sentences within the text. The following are marked (Table 32.1):

Table 32.1

Long-term processes	H	High intonational phrase boundary
	L	Low intonational phrase boundary
Mid-term processes	T+	Turn-up of pitch
	T–	Turn-down of pitch
Lexical markers	S	Stressed syllable
	U	Unstressed syllable
Modifier	F	Overlays pitch accent and expressive content

- sentence and intonational phrase boundaries–marked H (high) and L (low)
- smaller domains, syntactic phrases, defined within intonational phrases–marked T+ (turn-up) and T– (turn-down)
- lexical stress for each syllable–marked S (stressed) and U (unstressed)
- sentence focus–marked F (focus) on a single syllable.

F is a device for introducing simple modifications to an otherwise neutral intonation. In our description here we bring this out to indicate the basis of the mechanism for varying intonation for expressive and other effects–though, clearly, this single example marker would not be sufficient (Tatham and Morton 2004). Although F is tied to a particular syllable in the SPRUCE implementation described here, it constitutes in principle an element within the generalised prosodic wrapper. Early versions of the T&M model wrongly equate S and U with the H and L tones to be found in ToBI. ToBI uses H and L to characterise pitch accent, whereas T&M use S and U as markers of lexical stress.

T&M recognise in the model three *types of process* derived from the theory of cognitive phonetics:

1 Linguistically irrelevant *incidental processes.* These have mechanical or aerodynamic sources and do not contribute relevant information. They are independent of language and are void of linguistic information. Although often considered to introduce noise into the system, they are perceptually relevant because, while not linguistically systematic, they are nevertheless expected by the listener. They contribute to cueing that the speech is human, and are therefore essential in synthesis. Examples are breathing and other aerodynamic or mechanical processes associated with the overall mechanism involved in speech production.

2 *Phonetic processes* intrinsic to the rendering system and which can be manipulated under cognitive control. This is the process widely known as 'cognitive intervention'. Although the manipulation takes place at the phonetic level, they are nevertheless phonologically relevant; and in the production of an utterance these processes are carefully monitored and supervised. An example here might be the cognitive intervention into the aerodynamic processes which are responsible for voice onset timing to produce linguistically relevant variations, such as the contrast in English between initial [+voice] and [–voice] plosives. Cognitive intervention is a managed or supervised process played out across the entire utterance scenario.

Table 32.2

Tier	Information source
Long-term	Sub-glottal air pressure progressive fall–a linguistically irrelevant process, assumed to give a progressive fall in fundamental frequency
Mid-term	Intrinsic rendering process responsible for controlling the inclination or declination of air pressure, increasing or decreasing fundamental frequency (turn-up and turn-down)
Short-term	Manipulation of vocal cord tension, giving local changes in fundamental frequency

3 *Phonological processes* extrinsic to the rendering system. These are deliberately manipulated, though subconsciously, and produce phonologically relevant effects. Although cognitively sourced and not subject to the kind of external deformation expected of phonetic processes. These processes are still monitored and supervised in the production of an utterance.

The T&M intonation model adapted for SPRUCE synthesis is hierarchically organised, with a tiered structure reflecting long-, mid- and short-term information (Table 32.2).

The implementation focuses on

- a philosophy of integration of phonetic, linguistic and computational elements;
- the establishing of detailed and explicit relationships between physical and linguistic representations–this contrasts with most models which focus on linguistic representations;
- an entirely knowledge-based approach focussing sharply on the explicit manipulation of tiered data structures;
- extensibility directed toward incorporating variability, pragmatic and expressive effects.

33

Phonological and Phonetic Models of Intonation

33.1 Phonological Models

At their simplest, phonologically based models are designed to explain intonational processes underlying the phonetic rendering of prosody. Phonetically based models attempt to explain intonational processes involved in rendering the requirements of a phonologically derived utterance plan. Phonological models may point to particular, but later, phonetic processes, and phonetic models may point to underlying, earlier, phonological processes–where 'later' and 'earlier' describe logical rather than temporal relationships. Critical to the way in which the levels are related (unless this is made explicit in the underlying model, as in the case of the T&M model of prosody) is the pivotal nature of the markup system, which offers the opportunity to bring together the waveform and the associated symbolic representation.

By definition, phonological models are within the linguistics tradition, and attempt to explain the intonation system of a language. In doing so they set out the principles of intonation as exemplified in the relationship between intonation processes and the linguistic units of intonation: syllables for the most part. As such, phonological models are not seen as directly related to the physical waveform. As analysis tools, though, phonological models (e.g. ToBI) relate to the analysis of waveforms–a process which can be more or less explicit depending on the particular approach. Phonological analysis of prosody usually has some rules, but often relies on the intuitions of those engaged in transcribing or marking up the waveform using a symbolic representation.

33.2 Phonetic Models

Phonetic models, in contrast, focus much more on how phonetic intonation events align with phonetic segments, in as much as they can be located–either as temporally ill-defined stretches of signal, or as delimited stretches. There are theoretical problems here associated with trying to relate directly a symbolic representation with a physical representation, but models which recognise the problems and are careful to establish conventions of relationship do help solve the practical problem, if not the theoretical one.

Developments in Speech Synthesis   Mark Tatham and Katherine Morton
© 2005 John Wiley & Sons, Ltd.   ISBN: 0-470-85538-X

33.3 Naturalness

To advance speech synthesis to include good rendering of naturalness we need models which are designed to explain not just basic (some would say 'neutral') prosody, but also expressive content, as communicated (deliberately) or revealed (involuntarily) by the prosodic features of the soundwave. It goes without saying that, for the transparency that arises from adequate compatibility, the underlying phonological prosodic scheme has also to be explained, and preferably in the same terms as the phonetic approach to explaining the rendering process. The underlying linguistic structures which can be characterised in a nonlinear hierarchical model contribute unarguably to the robustness of any attempt to produce an intonational model–or a general prosody model–for synthesis. The key is in the mapping between underlying abstract representations and their associated processes, and derived representations and their more surface physical processes, and the final soundwave. The chain of events is shown in the diagram; the focus is on the pivotal plan in the middle.

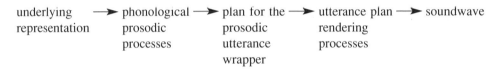

The phonological labelling task involves marking either tone targets or shapes–that is, interpolated tone targets. In general these are aligned with syllable nuclei which, despite the mismatch between abstract symbols and physical waveforms, are easily defined as the midpoint of the phonetic vowel nucleus. Figure 33.1 shows some sample waveforms with the syllable midpoint marked approximately. Since alignment is with the *syllable*–a quantitatively imprecise unit–the exact position for marking, were it theoretically possible to find one, is not too important.

In automatic systems there are problems with determining what is happening at syllable margins, and even with hand marking judgement is sometimes difficult or ambiguous. Some of the margin activity in the waveform is associated with local fundamental frequency perturbations, sometimes confusingly called micro-intonation. These effects are the product of aerodynamic coarticulation and have nothing to do with prosodics. Another local variation of fundamental frequency–this time spreading out from the 'centre' of the syllable–is found in the tones of many languages. This is a segmental use of fundamental frequency contour and, once again, not a prosodic effect. These examples illustrate a repeated message in this book when discussing prosody: the parameters of prosody, whether phonological, phonetic or acoustic, are multi-purpose carriers. This fact makes it especially difficult to analyse prosody into all its functions, and just as difficult to synthesise it.

Because of its symbolic high-level nature, phonological labelling can be a little vague in terms of identifying boundaries between segments. But the same is not true of symbolic low-level phonetic labelling. This is also symbolic, but relies on the notion of boundaries

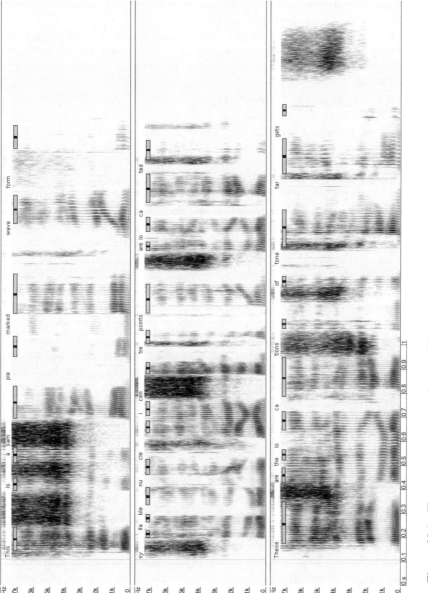

Figure 33.1   Three utterances, top to bottom: *This is a sample marked waveform. Syllable nuclei centre points are located. These are the locations of tone targets.* The shaded rectangles toward the top of the spectrograms indicate each syllable nucleus—the obligatory vowel which must be present in any syllable. The centre point of each nucleus is marked by a black area within the box.

and especially on alignment with acoustic boundaries. The symbolic representation uses elements which we call *intrinsic allophones*, and these are sufficiently 'close' to the waveform to require noting phenomena like the rhythmic changes involved in signalling, say, focus or emphasis which may involve moving boundaries. In unit selection synthesis, systems boundary marking of sound units within the waveform is essential for the accurate excision of units, but even in these systems prosodic label alignment is less critical. Even if phonological models of intonation can make good use of discrete labels without too many theoretical problems, phonetic models need to overlay on to segmental representations continuous representations of dynamic physical phenomena (fundamental frequency, durational and intensity movement).

This has worried some researchers, notably Taylor (2000) and Ladd (1996), because it violates the usual stance in linguistics to use only discrete symbolic elements in its representations. Taylor's compromise is that the prosodic phenomenon–without considering its phonological or phonetic analysis–is continuous by its very nature, so can legitimately within linguistics fall back on a continuous representation. We do not see this as a problem, provided it is realised that the 'horizontal' continuousness of prosody is a phenomenon associated, in the physical world, with clock time, and in the abstract world with notional time. *All* models preserve linguistic discreteness in their 'vertical' aspect by allowing distinct contours. Some of the distinction between contours in this vertical aspect is linguistic in nature and some reflects a natural variability. Interestingly, just as with segments, two different or varying objects may be linguistically significant on one language or dialect but not in some other–and when they are linguistically distinct they may be so at a phonemic or an extrinsic allophonic level. The phonemic and allophonic concepts have a place in distinguishing vertically between otherwise continuous prosodic contours.

The continuousness of prosodic contours is an attribute dictated by the nature of the acoustic rendering. But this continuousness is not necessarily exactly the same as the continuousness which coarticulation theory imposes on discrete segments. We may (the best example is ToBI) have an underlying representation which is in terms of–for the acoustic signal–widely spaced tonal elements; but it is not necessary to go on to posit a kind of coarticulation which joins these together in the physical world. It is better to say, quite simply, that discrete is rendered as continuous, and leave it at that.

The *real* problem which the continuousness idea obscures lies in the vertical dimension. Segmental variability, whether its source is linguistically relevant or not, can be characterised as the simultaneous existence of multiple allophones from which one is relevant at any one moment. *Which* one is selected by hierarchical context which may be phonological or phonetic (or, in the case of cognitive phonetic processes, span the two). In the case of prosody, however, the linguistic exponent is not a string of physical segments (albeit coarticulated), but a continuous curve selected from a set of possible discrete simultaneous curves. We have to be careful here of the problem we have mentioned several times: an *exemplar* derivation–segmental or prosodic curve–is not an *actual* instantiation, but one of many simultaneously existing objects. Although we generally focus on the continuousness of prosodic, especially intonational, representations, it is reasonable to transfer that focus to the vertical discreteness we have just mentioned. This brings the focus of prosody more in line with segmental modelling, and certainly enables us to avoid thinking of prosody as something special and perhaps a little less linguistic than segments. Prosody may appear to be more tied to the physical signal, but that, to us, appears to be an artefact of the descriptive model.

Certainly phonological intonation models like ToBI (with its surface, linear approach) and T&M (with its hierarchical, nonlinear approach) are properly linguistic and are couched as explanations of

- intonational structure (internal logical relationships–ToBI (surface) and T&M (deep))
- surface outcome (utterance plan plain prosodic wrapper–ToBI and T&M)
- pragmatic outcomes (utterance plan complex prosodic wrapper including expressive content–explicit in T&M)
- perceptual outcomes (predominantly implicit in ToBI, explicit in T&M).

33.4 Intonation Modelling: Levels of Representation

Jassem (2003) proposes that in a model of human intonation it is useful to recognise four distinct levels of representation.

1 *Acoustic* involves the extraction of the fundamental frequency as a time function.
2 *Acoustic phonetic* involves the fundamental frequency curve being 'smoothed and normalised for the speaker's individual mean pitch and range–essentially removing personal variability' (p. 294).
3 *Perceptual phonetic*–otherwise called *surface phonetic*–is an example is the 'expert's inter-linear impressionistic tonetic transcription' essentially trying to graphically represent impressions of intonation using a musical score type of notation. It is important to note that the expert is not graphing an impression of fundamental frequency but of intonation–that is, a *perceived* linguistic object assigned to *heard* fundamental frequency.
4 *Phonological* involves the underlying linguistic structure associated ultimately with the acoustic signal, but also the perceptual phonetic representation.

For the purposes of first-approximation modelling of intonation, the most important acoustic parameter to be isolated is clearly fundamental frequency where this is available from the periodic sections of the signal. Where no periodic signal is present, the concept of fundamental frequency makes no sense. However, that does not mean that we do not need to have at least durational information about aperiodic or silent sections of the waveform, since listeners are known to tunnel through these sections when they assign intonation cognitively. This is not unlike the general problem of characterising the perceived auditory scene and how it differs from the actual acoustic signal (Bregman 1990; Cooke and Ellis 2001).

Whether or not the acoustic phonetic representation should rely on smoothing and normalisation is arguable, unless this is very carefully done in the light of a comprehensive model of variability. For example, if subtle variations in the signal are there because they encode expressive content they will be lost, thus reducing the model's power. Since there is more call these days to improve our knowledge of expressive content and how it is encoded on to the parameters of intonation, a great deal of thought needs to go into any smoothing or normalisation process. One suggestion is to follow Jassem's general approach and derive a fully smoothed curve, using this as the basis for investigating what has been removed by simply subtracting it from the original curve. This will not guarantee, of course, that the 'right' variability has been removed, or indeed all of it. The situation is further complicated by the fact that, if perception is a knowledge-based system, the signal probably contains no more than triggers, or pointers to the listener's stored information.

Jassem is distinguishing between acoustic phonetic and perceptual phonetic levels in terms of bridging the gap between the physical signal (fundamental frequency contour) and abstract symbolic representations (intonation). Jassem recognises the levels, and the symbolic representation needed for perceptual phonetic representation; but ultimately what is at issue here is the way we characterise how impressions of intonation actually relate to the quantifiable signal. The problem is not less complex for speech synthesis than it is for automatic speech recognition, since it is this characterisation which forms the basis of conversion from (in our terms) the abstract representations of phonological prosody in the utterance plan and the actual soundwave to be created by the low-level synthesiser.

For Jassem, the acoustic phonetic representation is a kind of inter-level between the signal and his perceptual phonetic representation; and provided that precautions to protect information bearing variability are in place, the level can be derived automatically. It is a much larger step, though, to derive the perceptual phonetic level automatically. We come back here to our warning that these processes of data reduction are essentially lossy, and that recovery of the lost information–essential for natural synthetic speech–is highly problematical. Jassem projects a plan of research to develop the model, and it is clear that he has isolated important areas which are critical in terms of how they relate.

Part IX

Approaches to Natural-Sounding Synthesis

34

The General Approach

We find that researchers have modelled expressive content and emotion, broadly speaking, in two ways–by category or by process. Many researchers combine the two approaches, working with category constructs operating within process systems. The term 'process system' has been used to refer to models that simulate natural processes; it involves stating ongoing changes in the system, along with specifying structures that will facilitate this change. The term *process* system has a technical meaning in computer science which may be used less technically in other disciplines (see Chapter 26) This approach is in contrast with a black box model, employed for decades particularly in psychology (Wehrle and Scherer 2001), and which is designed to produce outputs that are similar to the output of a natural system. In black box modelling, *how* the output is produced is not of importance; mapping from input to output is adequate.

Not everyone follows a narrow definition of process modelling. But the following outlines some of the differences we have found useful to note between category and process, in as much those words appear to be used by researchers in expression and emotion studies.

Differences in the approaches can be seen using an analogy from phonetics.

- One approach observes that the tongue changes shape. This can be seen on a series of x-ray snapshots–an x-ray movie. The *idea* of change in time is supplied by the person looking at the x-ray (category), and remains *implicit* and hidden in the model.
- The other approach looks at the half dozen muscles that move the tongue. These muscles are at different positions in a three-dimensional space and interact to give the overall shape to the tongue. Dynamically varying the shape of the tongue produces different articulations resulting *via* aerodynamic processes in a signal perceived as a sequence of sounds by the listener. Time is *explicit* in the movement of the muscles (process) and overt in the model.

34.1 Parameterisation

Suppose an emotion can be viewed as an amalgam of several underlying parameters. Initially the underlying biological properties result in tension in all muscles, including those which innervate the vocal tract. These effects may be experienced and reported by the individual as *emotion*. There may be cognitive underlying features, perhaps as a result of

Developments in Speech Synthesis   Mark Tatham and Katherine Morton
© 2005 John Wiley & Sons, Ltd.   ISBN: 0-470-85538-X

appraisal in which an event is evaluated as to being good/bad for the individual and which contributes to his or her response. If these underlying processes are

- parameterised
- correlated with features in the speech waveform
- recruited to account for differing emotive effects by reason of their varying combinations of parameter values

then a way of accounting for different emotions formally and predictably is possible. A parametric approach provides a framework for comparative evaluation and predictability based on relative parameter changes. Above all, different phenomena–in our case, different emotions or forms of expression–are seen to have commonality in their sharing of the same set of underlying parameters, but difference in the values each have for the parameters.

Using linguistic descriptions, it is possible to map from a phonological specification to phonetic descriptions underlying the soundwave, and to correlate these descriptions with descriptions in parameter sets in a different domain. For example, in the early JSRU system (Holmes *et al.* 1964) we find static descriptions which characterise linguistic input to the synthesiser software in the form of parametrically organised tables of values. These abstract representations are transferred to the synthesiser, itself operating parametrically but in the physical world. The result is to establish a transparent relationship between the abstract linguistic units and to relate them, equally transparently, to the physical synthesis process. The importance of this formal and explicit theoretical breakthrough in synthesis studies should not be underestimated. Other researchers in synthesis came up with the same idea around the same time, but a notable precursor of the general idea can be found in Lawrence's 'parametric analytic talker' (Lawrence 1953).

For emotive content, a higher level linguistic and expressive/emotive specification, along similar lines, might also positively correlate with acoustic parameters. Listener response can also be modelled in terms of evaluating the speech output parameters, either the obvious acoustic triggers or those inferred, to conclude of the speaker that, for example, *They are happy* or *They are angry*, etc.

34.2 Proposal for a Model to Support Synthesis

The basis of our proposed model is the two-planes approach involving separate, but transparently related, descriptions of both static and dynamic properties of language. We have seen that traditionally transformational generative grammar has analogously separated knowledge from knowledge usage, while characterising explicitly only a database of the static properties of language (Jakendoff 2002; Hayes 1995; Goldsmith 1976; Hyde 2004; among others). The basis for modern phonetic theory and the kind of theory needed to support speech synthesis (and automatic speech recognition) has, however, to refocus this traditional approach and give equally explicit support to characterising dynamic processes in language. In our case we are interested for the most part in speech production and perception, and take as read details of the underlying processes of syntax, semantics etc. We also need to include less well developed areas of language, including especially pragmatic considerations such as relevance, inference and implication (Grundy 2000; Levinson 1985; Kearns 2000).

34.3 Segments and Prosodics: Hierarchical Ordering

We want to put together a set of declarative statements describing speech production. The standard minimum description consists of two main types of object, segmental elements and prosodic elements. Using XML we can declare two possible relationships between them:

```
<segmental_system>
        <prosodic_system/>
</segmental_system>
```

or

```
<prosodic_system>
        <segmental_system/>
</prosodic_system>
```

1 In the first model, the phonetic description of the segmental system is basic, with prosodic features added later–that is, after the segment string has been characterised. For example: a syllable (describing a group of individual segments) may subsequently have attached to it a prosodic feature, such as intonation change. In phonetic terms, a phonetic grouping identifying this syllable would have an added changing fundamental frequency associated with it when instantiated. In other words the focus of the approach is the segment; once strings of these are available they are given a prosodic contour. We could consider the utterance *John's gone home*, and the possibility of saying that this could be given either a falling intonation contour (to identify it as a statement) or a rising intonation (to identify it as a question). The utterance pre-exists without prosody, and prosodic features are *added* to it since they are specified lower down the hierarchy.

2 In the second model, a prosodic or suprasegmental framework is specified first, and the segmental representations of the utterance later *fitted* to it. In this model, prosodic features are available as an environment within which the utterance is formulated. An example might be an overall decision to speak fast before any thought is finalised; but when it is finalised it is encoded within the context of being uttered fast–not formulated as neutral and then subsequently modified when, as in the first model, the prosodics are 'discovered'. In the second model prosody dominates irrespective of any one utterance which may be fitted to it. Taking the *John's gone home* example, we could hypothesise an internal dialogue for the speaker like: *I am going to ask a question, and my question is: John's gone home?* In model 1 this would take the form *John's gone home, and let me put this as a question: John's gone home?*

The surface result of either approach may be the same–we get the perception of a question about whether John has gone home. But we see clear advantages of such an approach, when, for example, structuring a dialogue. There are also clear advantages when prosody is used to render voluntary expressive content: *I have decided to be purposeful, so now let me say what's going to happen.* Here everything the speaker says is going to be *coloured* by purposefulness, and this expressive content becomes all-pervasive because it *dominates* everything until further notice. We will have more to say about this later.

In terms of speech synthesis there are no overwhelmingly compelling isolated reasons for choosing the first model over the second–though it does seem that in psychological and linguistic terms model 2 is to be preferred. However, there are one or two observations which, in our view, tilt us toward preferring model 2 for synthesis.

1 No human speech is ever produced without prosodics–all utterances must have a prosodic contour.
2 Prosody extends beyond the sentence domain, and this feature is particularly prominent in paragraphs within a single speaker, or dialogues when two or more speakers are involved.
3 All human speech has expressive content which is encoded using prosody.
4 Expressive content often goes well beyond sentences, and in dialogues can stay with a speaker even when that person has been interrupted by someone else (or a machine).
5 There seems to be no empirical evidence that speakers produce sentences in two stages, a segmental one and a prosodic one, which are actively blended to produce a final result.

We feel that for speech synthesis this collection of observations is sufficient for us to opt for the model which provides for the planning and rendering of utterances within an overarching prosodic framework. In other words we have a preference for formally setting up a prosodic wrapper within which to work toward individual sentences. The approach will enable us to carry over the prosodic message (the pragmatics and some of the utterances' semantics) from utterance to utterance, setting the scene for a speaker's delivery.

34.4 A Sample Wrapping in XML

Let us consider a small example of the wrapping hierarchy. An utterance can be regarded as a sequence of syllables structured in a particular way to express what is called 'segmental meaning'–that is, the overall meaning associated with the words chosen for the utterance. The semantics and syntax underlying the utterance develop a hierarchical relationship, the basis of which is the sentence (though this does not preclude the possibility of overarching domains like *paragraph*). In terms of utterance prosody, our hierarchically structured arrangement of syllables has to be fitted within a pre-determined prosodic structure, itself hierarchical for intonation. The overall structure looks like this:

```
<prosody>
        <IP>
            <AG>
                <rhythmic_unit>
                    <syllable number="1-4"/>
                </rhythmic_unit>
            </AG>
        </IP>
</prosody>
```

We characterise the syllable within this structure as

```
<syllable>
        <onset>
                <consonant number="0-3"/>
        </onset>
        <rhyme>
                <nucleus><vowel number="1"/></nucleus>
                <coda><consonant number="0-4"/></coda>
        </rhyme>
</syllable>
```

The intonational structure contains, within this example, a sequence of up to four syllables–themselves arranged in a hierarchical structure (not shown). Intonation is seen as an arrangement of accent groups (AG) which include rhythmic units, the whole grouped into intonational phrases (IP) which form part of the prosody. Our model defines a part of intonational prosody, *not* the intonation of a particular utterance; the hierarchy is completely general and requires, as part of the rendering/instantiation processes, an utterance to be fitted within it.

It is possible, and common practice in linguistics, to characterise prosodics independently of sentence and cognitive expression, and of biologically sourced emotive content. These characterisations lay out the potential of the prosodics in an abstract way–what the units are, how they can combine, etc. In the model we are suggesting here, the prosody needs also to indicate points at which there can be interaction with other parts of the model. In Part VIII, we incorporated this basic structure as part of a larger one wrapped in expression.

34.5 A Prosodic Wrapper for XML

Characterising the sound pattern of plain messages occurs on the phonological static plane. Some linguistic expression can be conveyed which is not necessarily emotive, but we suggest characterising this as a subset of the phonological expressive/emotive content module. This module will characterise intended cognitive input such as 'emphasis', a linguistic feature which does not appear to have biologically sourced emotion. For example, emphasis–sometimes referred to as 'prominence' or 'contrast'–can be found in sentences such as

- *prominence* as in: *No, I said I* <u>don't</u> *care for strawberry; I'd like chocolate*; or
- *contrastive* emphasis such as: *I said* <u>brown</u> *not* <u>crown</u>.

These messages are intended to correct information that seems to have been misunderstood, not *necessarily* any emotion at all–a linguistic function rather than an expressive one.

We suggest that all processes, whether a general type characterised on the static plane or a specific type characterised on the dynamic plane, fit within a prosodic wrapper. In the static model, planning takes place within a prosodic framework which constrains or puts boundaries on planning; rendering also occurs within a prosodic framework (but with a different status: a phonetic prosody). We need two wrappers–one for phonological planning and one for phonetics within which actual production occurs.

The phonetics specifies the configuration of the vocal tract which is the means for producing the acoustic carrier, and which is affected by the total biological stance of the individual.

34.6 The Phonological Prosodic Framework

The phonological prosody framework reflects paralinguistic phenomena such as cognitively sourced expression, and pragmatically derived effects. These last could have an influence on phonological choice such as contrastive emphasis in the phrase *No, I said the 'white house, not the 'White 'House*. The complexity here is compounded because of the contrastive emphasis on *'white house*, opposing it to *'black house*, for example. Non-contrastive *white house* and *White House* differ in another way.

Dynamic information channels which call for specific phonological or phonetics effects call for them at the time of producing the utterance; they form an *additional* layer of choice. That is, they are not part of the set of statements which characterise the general linguistic structure of potentially all sentences and utterances on the static plane. These choices are of the moment, are temporally constrained by the conversation context. We suggest these channels of information feed into the production of a straight linguistic substrate as in Figure 34.1. The phonetic prosodic wrapper takes information from the phonology, contained within the phonological prosodic wrapper, and renders this specification according to information on the phonetic static plane.

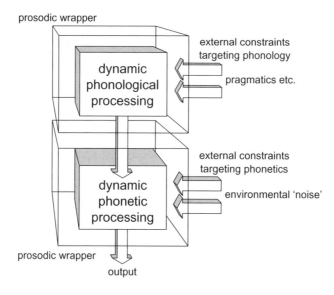

Figure 34.1 Simplified diagram of the dynamic section of the model. Incoming constraint channels target both phonological and phonetic processing, distorting the prosodic wrappers.

35

The Expression Wrapper in XML

We suggest that expression itself wraps the entire basic prosodics envelope. We want to characterise expressive/emotive speech as it can be conveyed to the listener by the vocal effects described by linguistics as *prosodics*.

Current synthesis systems *add* emotive effects to a speech synthesis system. The standard method has been to produce a synthetic utterance with no deliberate expressive content (so-called 'neutral' speech), and then manipulate the prosodic, or suprasegmental, features to produce speech judged to have expression/emotive content.

Using XML notation, this can be sketched in the standard model as

```
<linguistic_output>
        <expression_emotion/>
</linguistic_output>
```

That is, expression is lower in the data structure hierarchy than the linguistics. We suggest another way:

> Regard expressive/emotion content as the major node; and regard prosodic information as contained within this node; and, in turn, segment descriptions within the prosodics.

This can be sketched as

```
<expression_emotion>
        <linguistic_output/>
</expression_emotion>
```

We have chosen this second possibility and base the choice on the following considerations.

1 In writing generative grammars, prosodics occurs prior to the development of phonological segments, and phonetic segments. If grammars reflect our intuition as to the elements and structure of language, this might be evidence for our description of prosodics occurring before segment specification–or 'wrapping' the segments in prosodic features. And if emotive content is expressive, and if expression is carried by prosodics, then we can say that, using XML notation.

Developments in Speech Synthesis   Mark Tatham and Katherine Morton
© 2005 John Wiley & Sons, Ltd.   ISBN: 0-470-85538-X

```
<expression>
        <prosodics>
                <segmental_representation/>
        </prosodics>
</expression>
```

2 Changes in prosodic features during speaking are meaningful to various degrees, and therefore they need to be accounted for.

3 There is no convincing argument for the alternative approach of producing a sentence cognitively and subsequently adding expressive/emotive content while speaking. The formalism in XML allows us to declare that expression wraps prosodics, and the prosodic features wrap utterances. That is, the utterance is seen as produced within the overall pattern of the intended expressive content.

In a sense, therefore, information from the sentence module and information from the cognitive expressive/emotive content module are both available to the phonological prosodic module. It is here that 'basic' specifications are modified by information from the sentence and expressive/emotive content modules–each higher in the tree.

35.1 Expression Wrapping the *Entire* Utterance

Consider a general framework for expression, suggesting in outline some structural equivalence of several different modalities (writing, speech, dance). It might be sketched (for expression) as

```
<expression>
        <writing/>
        <speech/>
        <dance/>
</expression>
```

and (for message) as

```
<message>
        <writing/>
        <speech/>
        <sign_language/>
</message>
```

with an expansion for writing as

```
<writing>
        <alphabet_system/>
        <ideograph_system/>
        <hieroglyph_system/>
</writing>
```

Using this kind of hierarchical approach, the general framework we are suggesting as a model of expression using linguistic units only with expression wrapping prosody is

```
<expression>
        <writing/>
        <dancing/>
        <speaking>
                <prosody>
                        <IP>
                                <AG>
                                        <rhythmic_unit>
                                                <syllable number="1-4">
                                                        <onset/>
                                                        <rhyme>
                                                                <vowel_nucleus/>
                                                                        <coda/>
                                                        </rhyme>
                                                </syllable>
                                        </rhythmic unit>
                                </AG>
                        </IP>
                </prosody>
        </speaking>
</expression>
```

where we give just one of each of the elements IP, AG, rhythmic_unit and syllable, though of course the actual number with vary with the utterance between the prescribed limits.

In this way it is possible to build up hierarchical data structures, of which the above are just simple examples, which incorporate several modalities–each of which can be related or used in the same scenario. Thus expression in the same human being can *simultaneously* dominate the way he or she writes, dances, speaks etc. The expression can be regarded as one of the unifying elements within the generalised data structure. We could also imagine that *rhythm*, as an element, can occur under *speaking* as well as under *dancing*. Inter-modality relationships have not yet been called for on any large scale in simulations of human behaviour, but there can be no doubt that such a requirement will arise in the not too distant future. The relationships exist side by side in human behaviour; it makes sense to try to introduce a unifying formalism to capture aspects of these relationships.

35.2 Sourcing for Synthesis

Whether expression is cognitively or biologically sourced is not important for synthesis unless the origin is important for the synthesis model. The way anger is modelled may depend on the physical and/or cognitive state–the detail of the acoustic signal might differ depending on the source. For example, raising fundamental frequency as a global feature may have a different effect on the speech output than increasing the fundamental frequency on a single word–and the first case may result from an overall biological stance. The second might be a function of cognitively selecting a particular word, or from cognitively intervening in the stance, thus changing the vocal tract configuration briefly for the production of that one word.

In the general theory of speech production, considerations like this are very important, but it is easy to argue that they are less critical in any theory designed to support speech synthesis or automatic speech recognition. There are a number of questions which need answering to settle what must be or need not be included. What follows is just a sample.

1 Does inclusion of this or that part of speech theory improve the overall synthetic output? Here we're concerned with features which lead to a big gain in, for example, naturalness– the inclusion of prosodic variability, for example.
2 Does inclusion make it easier to include other more important features? For example, if you include a proper model of prosodic variability it is much easier to add expressive content.
3 Is the speech synthesis to be used in research experiments designed to reveal how human beings behave? Researching how people perceive can often benefit from synthetic stimuli in experiments because they are more manipulable than human speech. But if the synthesis is not faithful to human speech or how it is produced, then the results might be about the perception of synthetic speech rather than the perception of real speech.
4 Can inclusion, while perhaps less important now, lead to easy and predictable extensibility of the synthesis system in the future? Two obvious examples here are expression (just commented on) and extensible inclusion of dialect/accent. Adding accents to a general system in the 'wrong' way might make the system quite unwieldy.
5 How much interaction is there with human beings? Dialogue systems are better if they seem to 'understand' the people they are interacting with.

35.3 Attributes Versus Elements

In describing expression in XML, the question arises as to differentiating between what features are properly elements and what are properly attributes on elements. Often this will be obvious, as when characterising degrees of expression–degree seems to fit easily with the attribute concept. But sometimes things will be less obvious. So, for example, irritation might be a degree of anger:

```
<anger degree="2">
```

where the range is 1–5, with severe anger at "5" and somewhat irritated at "1". And ecstasy might be a degree of happiness:

```
<happy degree="5">
```

where the range is 1–5, with ordinary happiness at "3" and ecstasy at "5". Or they might be treated as separate elements:

```
<expression>
      <anger/>
      <irritation/>
```

```
        <happy/>
        <ecstatic/> ·
    </expression>
```

At its simplest we might just ask the questions:

1 How deep do we want the hierarchy to go?
2 What is to be gained by having *anger* on a par with *irritation*?
3 Is it important to show transparently that *irritation* is a type of *anger* and not dissociated
  from it as a class (i.e. a member of a different class of objects)?

In a less trivial way we may be concerned to show that there is, for example, a class
of expressions called *anger* and that there are various degrees or sub-types of this that are
rendered, not as different processes, but as sub-types or as variants of a single process. This
would be a *real* gain in the model building process.

Another approach might involve assessing the actual biological or cognitive source of
expression: are two surface exponents of expression (e.g. *anger* and *irritation*) derived
from the same basic source? If the answer is yes, it makes sense to show them as basically
the same element but with accompanying and distinguishing attributes. The problem is
the way we label these features of expression in everyday language. It is very common
to focus on different lay words to distinguish between what in fact are degrees of the
same phenomenon–and, of course, vice versa. If this is a mistake in lay language–and of
course it is *not*: it is just the way we do things–we must not commit the same error in our
science.

Yet another consideration is whether symmetry in the model is productive. Thus, for
example, would it be useful to think of all types of expression as being able to manifest
themselves to listeners in terms of degree? If so, degree and the *way* degree as an attribute
is actually used becomes paramount in the characterisation. Such questions are not trivial,
and their answers are not always obvious. Two major factors can be appealed to.

1 Is the approach productive?
2 Does the approach reflect what we know of human beings?

When constructing a dynamic model using attribute values which change with time,
the values on the attributes and *relation* between attributes must be carefully specified. The
values of attributes can be relatively easily specified in XML, but the *relation between attributes*
seems to be more difficult. Relationships between elements are quite easy to establish–the
explicit hierarchy of the overall data structure should be designed to take care of this. But
how attributes relate to one another is a different problem because by simply listing
attributes (see numerous examples in the SSML recommendations, commented in Part V)
the transparency of any hierarchy is lost. The problem could be solved by 'promoting' attributes
to element status–but then the advantages listed above would be lost.

There are some important questions concerning relationships between elements, between
attributes, and between elements and attributes, which should be considered so that the
assigning of element or attribute status does not become random in the model. So, among
other things, we might ask:

1 What would be the optimum way of characterising relationships within our XML data structure?

2 What kind of relationship *can* be specified, and would this meet our needs?

3 Can these types of relationship be used in characterising the underlying structure of synthetic speech, particularly in the area of contributions to naturalness?

4 Under what conditions would the relationships hold, and what happens as the conditions change?

5 Under what conditions could the relation *not* hold? That is, what are the exceptions which need to be transparently characterised in the data structure?

6 How could the specification of elements and/or attributes be formulated so as to form a hook which could be accessed by a procedure? This is important in the rendering stages where data structures may need some procedural processing.

There is one particularly important property of prosodic features we need to consider carefully when discussing the relationship between elements and attributes.

Most prosodic features vary with time whether they are used to characterise basic prosodics or whether they are being used as the vehicle for expressive content. Put another way: prosody and expression are *time-varying*.

Exactly how to incorporate time-varying conditions into XML markup is by no means obvious. In fact it becomes fairly clear that XML markup is more suited to characterising static data structures than it is to varying data structures. We have already mentioned how the SSML team try to characterise time-varying fundamental frequency contours–by a kind of halfway house between a static description and a full-blown dynamic description, marking relevant turning points in the contour and assigning values at these points, while establishing some meta-rule for joining the points. In practice this is rather clumsy, but theoretically it is useful in confirming a bridge between abstract and physical representations, provided there is an external set of conventions (the meta-rules) for interpreting the characterisation. If we take our cue from SSML we will try to include time variation in how we handle attributes (Tatham and Morton 2004).

Numerical representation supplies values for attributes, and these can be changing values, as we have seen. We have also seen that attributes can be given relative values–an abstract extension of the numerical approach. But we have also seen that in addition attributes can be used as descriptive modifiers in the manner of adverbs (qualifiers of verbs) or adjectives (modifiers of nouns).

35.4 Variation of Attribute Sources

In our model we need to incorporate physiological features of the vocal tract which may change under differing emotive effects which are *felt* by the speaker. But descriptions of biological or cognitive processes are not the same as a description of the appearance of emotive content in the *waveform*. It is possible to have attributes that derive from a biological module (described in an emotive module in the phonetic static model) or from a cognitive module (described in the expressive/emotive module in the phonological static model). Keeping these differences straight is not easy, though it is clearly necessary to distinguish between cognitive effects and biological effects, particularly if they enter into play at different stages in the unfolding of the model as phonology and phonetics.

For example, it may be useful to describe some aspect of emotive content as the basic biological stance taken in extreme anger where the muscular system becomes tense, and contrast this with the relaxed muscular system in happiness. In these two cases, the potential for tension and relaxation in the vocal tract might be related to the *biological* source of emotion–and to be characterised therefore within the phonetic component as affecting the way the acoustic signal is produced overall. Cognitive intervention may overcome effects of say, extreme anger, which might produce a different kind of tension than extreme anger uncontrolled by cognition. So an attribute might be 'capable of blending', 'capable of overriding automatic (learned as well) physical reaction'.

35.5 Sample Cognitive and Biological Components

The data structure in Table 35.1 characterises the biological source underlying acoustic properties of the signal which are detected and interpreted by the listener as anger on the part of the speaker. Mechanical distortions in the expected vocal tract configuration induce changes in the vocal tract that could appear as acoustic changes in the speech waveform.

Table 35.1 Sample cognitive and biological components

	Comment
`<expressive/emotive_content>`	
`<anger`	The emotion is anger, with attributes and values OR limits
`intensity=integer`	Intensity of emotion
`continuity=integer`	Duration of emotion
`modifiable="yes/no",` `integer`	Capable of cognitive intervention? Needs an integer value if degree is specified
`blendable="yes/no"`	Combine, blend, mix with other emotions
`detectable="yes/no"`	Reportable by speaker, awareness of emotion
`audible="yes/no"/>`	Detectable by listener
`<happiness/>`	With attribute list similar to *anger*, either with the same attributes, or different values or limits, or with different attributes that fit within the system and allow changes in the prosodic structure
`<fear/>`	Similarly
`<sadness/>`	Similarly
`</expressive/emotive_content>`	

In the characterisation this is a dynamic process, drawing on descriptions on the static plane in phonetics, and described by phonetic prosody. Continuous variation in the values of the attributes alter the acoustic characteristics as the conversation unfolds. The CPA monitors the ongoing situation and supervises as necessary.

35.5.1 Parameters of Expression

Thus parameters or features of some expression (e.g. 'intensity') will have their own intrinsic values, which together specify the given expression. In a simple model, the intrinsic values of all features together in a sense ARE the expression. But it may be the case that these features interact differently producing different expressive content. This might account for 'blends' (Tatham and Morton 2004).

35.5.2 Blends

The term *blends*, also called *secondary emotions*, refers to combining or merging the features of primary or basic emotions (Ekman 1992). If the association between four identifiable types of neural activity (Panksepp 1998) and the experimental inferences which label four basic emotions is plausible, then it is possible that varying the strength of the neural signal could result in varying degrees of reported emotion. One emotion parameter could, for example, be associated with a physiological vector labelled *intensity*. So, reporting irritation may relate to a low-intensity activation of a circuit capable of giving rise to rage if more fully active.

Reactions arising from four basic emotions of varying intensity, and resulting in large numbers of potential reported emotion experiences, would probably be limited by thresholding properties of the biological system. Limits might vary among individuals, leading to a large number of emotion types reflecting differing intensities and the individual's own sensitivity, awareness and appraisal. If more than one circuit is involved, with differing amounts of intensity and activation (and differing individual characteristics), the potential for many reported emotion experiences increases.

35.5.3 Identifying and Characterising Differences in Expression

A definition which asserts no more than that there is a conceptually isolatable difference between say, *pleasant* and *polite*, or *irritation* and *anger*, is not helpful to us. As speakers of the language, we know these words refer to differences. But researchers need a framework within which these differences can be described using as many different types of expression as possible so that they can be compared using similar principles and methods. We suggest a feature set, analogous with phonological feature sets, which uniquely identify all the objects in their domains. One feature, or parameter, could be *intensity*, indicating the strength of operation of this parameter under certain conditions. Parameters could have abstract names, names correlated with expressions or with reported emotive states, underlying cognitive or biological states.

For example, a set of intensity values from 1 to 5 could be marked 4 for *anger*, 1 for reported *irritation*; 4 for activity in the neural circuit associated with *anger* reaction, a low value for *irritation*. Low values for activity in designated circuitry indicate the possibility of lower intensity, but the reported emotion would depend on the action of other circuits and the influence of cognitive effects.

This type of characterisation could be used dynamically. Emotive content and modes of expression change in dialogue, and the period of time of this change varies considerably. Comments such as *He seems to be changing his mind* or *She sees this point I'm trying to make* reveal information which can be conveyed not by lexical selection or grammatical construction, but by a change in the tone of voice, and might be described by a set of changing features through the conversation. An analysis of the acoustic signal may not show an abrupt acoustic change; what is likely to have happened is that the combination of features on this occasion has resulted in the perception of a category change in the listener.

The task is to search for acoustic characteristics which might function as perceptual triggers or pointers. Some questions to ask are:

1 Is there a set of features which could characterise all modes of expression?
2 Should the values on the features be binary or multivariate?
3 Is it possible to predict the expression, given the set of features?

Some of the combinations would be prohibited–full strength for example on all four circuits would be unlikely. That is, is it possible to imagine extreme anger, fear, sadness and contentment occurring simultaneously? And is it difficult to imagine (even if occurring sequentially or blended in a time-space which does not allow recovery from one or more) what condition the individual would be in?

As an example, looked at from the point of view of the word used to describe an experience, such as *annoyed*, a possible contribution might be from *anger* and *fear*. The word *patience* could indicate the individual's attitude which might be composed of *contentment* with *irritation*, a very mild low-level anger. The concept of blends, of mixtures of more basic emotions, has proved useful in characterising the enormous range of reported emotion (Lazarus 1994) and suggestions about a hierarchical arrangement of such classification (Plutchik 1994). But if, as we suggest, biological and cognitive contributions to a synthesis model producing natural sounding speech can be productive, more work needs to be done in determining coherent modelling of these contributions suitable for implementation in synthesis systems.

35.5.4 A Grammar of Expressions

Going up a level, if individual expressions interact or combine with other expressions, we might write a *grammar* of expressions. For example, one particular expression may be able to be followed by or merge into some other expression but not with a different one. So expression E1 may be followed by E2, but not by E6. For example, anger may be able to dissolve into regret, but not into joy. There would be constraints (stated in the grammar) on what can occur–sequencing, blending etc. would all be affected (Tatham and Morton 2004).

36

Advantages of XML in Wrapping

It is clear that the biological source for some emotive or expressive content is distinct from the linguistic sources in the system. It is also clear that it is possible to identify whole areas of the system–in the case of the characterisation of anger above, the entire dynamic phonetics component–that are affected by these sources. It is also the case that cognitively sourced expressive content influences whole areas too–in the case of speech production, the entire dynamic phonological component. We can also see that cognitive expression also influences other areas of the linguistic system. A father, angry with son Bob for a particular prank, may address him as Robert rather than Bob, to underscore the seriousness of the situation, at the same time choosing a variant prosody to reinforce the point.

Since this content is comparatively all-pervasive in respect of identifiable groupings of processes in the linguistics, it makes sense to use a formalism which transparently makes this clear. Hence the XML wrapper assigned to emotive content–a wrapper which by definition is an element with a dominant effect in the hierarchy. Thus:

```
<cognitive_expression
        external_source = "no"
        prosodic_phonology_exponent = "yes"
        segmental_phological_exponent = integer>
                <dynamic_phonology>
                        <dynamic_phonetics/>
                </dynamic_phonology>
</cognitive_expression>
```

and

```
<biological_expression
        external_source = "yes"
        prosodic_phonetic_exponent = "yes"
        segmental_phonetic_exponent = integer>
                <dynamic_phonetics/>
</biological_expression>
```

The above two fragments of the characterisation of cognitively and biologically sourced expression respectively show attributes on the `<..._expression>` elements which contribute toward indicating the scope of the element in terms of its effect.

That `<cognitive_expression>` disturbs the way the prosodic phonology works is not the only influence. There is also an influence on the segmental phonology in the sense that particular segments may be selected (perhaps to change style or even accent) or particular optional rules not applied (as in the case of deliberate change in rate of delivery). This is shown by introducing the integer value on an attribute segmental_phonology_exponent which roughly shows the extent of the influence. `<biological_expression>` likewise disturbs phonetic_phonology as previously explained. But it also disturbs how segments are rendered. We include an attribute segmental_phonetic_exponent as a means of indicating the degree of influence exerted by the biological distortion.

Since we want to wrap the speaking in an expression frame, we need a formal way of expressing this. The formalism in XML allows us to declare that expression wraps the phonological prosodic units of speech–the intonational phrase, accent group, syllable, and segmental units of speech.

Our model must do at least the following:

1 Declare expression in general as the top layer wrapping the utterance framework.
2 Specify types of expression needed in speech that are linguistically sourced (including cognitive emotive content that can be associated with prosody).
3 Specify types of emotive expression that are cognitively sourced encoded in linguistic categories (words etc.).
4 Declare a phonological prosodic system that wraps sequenced (but hierarchically structured) syllables and segments, and can access on the static level the abstract specification of intended phonological units.
5 Declare the range of values on elements and attributes of prosodic characterisations with and without expressive and cognitive emotive content (i.e. abstract default and range of expressively dominated variability).
6 Declare a phonetic prosodic system that wraps phonetic syllables and states the abstract specification of the range of potential instantiations.
7 Specify a range of choices about required and optional features on the elements which may be used to characterise instantiations of expressive/emotive speech.
8 Declare the values (weightings) on elements in the static phonetics, specifying values in the possible range.
9 Specify types of emotive content which is biologically sourced and are encoded in the speaker's biological system.
10 Specify terminal nodes and their content type for segmental insertion.
11 State how weightings (values) might change *if* these are instantiated.

Our gaol is to describe what speech *is*, using a formal vehicle to characterise the data structure: XML. This type of structure allows us to specify what linguistic terms we are using to describe speech and what conditions allow choice, ranges of values (not the values themselves), required and optional objects, etc. The framework must be such that the source of non-linguistic content (cognitive, but not directly connected with language, and biological) can find an entry point into the data structure, and such that an appropriate response by the data structure is shown as a consequent potential modification of the data structure.

36.1 Constraints Imposed by the XML Descriptive System

In describing expression/emotion in XML, we ask whether the underlying characteristics of an emotion, for example, that *differentiate between* emotions can be described within an *attribute structure*. If so, can we use a universal set of attributes varying only with respect to their values, or would a more plausible model result if each emotion label required a set of specific attributes? This is distinct from the question of whether a particular property should be afforded the status of element or attribute. For example, a colour set is a universal from which a description can be drawn to uniquely define an object in terms of its colour.

XML seriously limits descriptions within a *dynamic model* when needing to use attribute values that might change with time. Clearly values on attributes and the relationships between them must be specified. The *values* can be specified in XML, but characterising the relationships between them seems to be more difficult. Some questions are:

1 Can relationships be specified in XML?
2 What kind of relationship can be specified?
3 Can these types be those that are need for expressive content?
4 How do we specify under what conditions identified relationships will hold?
5 Under what conditions could these relationships *not* hold? That is, what are the exceptions and are they *formal* or *substantive* in origin?
6 How could the specifications be treated as *hooks* accessible by procedures? That is, how do we use varying relationships to identify entry points into the data structures which all require procedural intervention–speech production is not simply a hierarchically declared data structure, it also has procedural features.

For the moment, it is useful to look at the emotive content and prosody relationship as one of constraint. The biological emotive content can be described as affecting rendering by seeking hooks to lock on to or to key into in the phonetic prosody module. For example, anger requires association with the prosody module and finds some unit (syllable, segment, phonetic IG) to lock on to; this is in the instantiation mode–the *potential* for this occurring is specified in the static modules, the *lockability potential.*

We have suggested that there is a prosody, a highly abstract potential capable of independent characterisation. This characterisation posits entry points, the hooks, where interaction is possible–from syntactically determined sentence or pragmatically determined expression, or both. One way of describing the locking in may be the use of the 'key' concept. *Receptor sites* within the prosodic model wait or *poll* for interaction attempts, while *elements* from sentence and expression *seek* suitable sites to key into. The receptor sites are coded as key-receptors; these may be unique or there may be a set of different types distributed around the prosody. Matching oppositely configured elements from sentence and expression seek key-receptors within the prosody to which they latch into if there is a key match.

36.2 Variability

We are very clear that enhancing the naturalness of synthesis by the inclusion of expressive or emotive content is extremely complex, and that earlier ideas that sought to model expressive content in the soundwave as a simple manipulation of a tiny number of parameters fell far short of the mark. Paradoxically the approach introduces distortions (like constant value expression, where expression is always a changing phenomenon) which actually worsens the judgement of naturalness. There are a number of important points still to consider.

1 If an emotion is strong (perhaps global), then many phonetic elements and attributes in phonetic prosody are affected. It may be the case that linguistic content is also affected.
2 Choice of words and their linguistic expression can be accompanied, with the speaker's awareness, by biological changes. These correlate in the model with changing phonetic prosodics, such as effects on the phonetic syllable or segment.
3 The level at which the key fits has to be determined. If the emotion is mild, then few elements and attributes in prosody are affected, or the key fits at a lower level.
4 It may also be the case that, if the emotion is mild, its intensity level is low, and this could be a *global* low-intensity marker at all the relevant levels. Another way of saying this is that there are local *vs* global considerations when it comes do characterising the domain over which expressive constraints operate.

A further possibility involves *thresholded* keys. Areas of the prosody may only come into play if the seeking key is strong. For example, if anger is intense, it may key into the rhythm section of prosody, but not if it is weak. We could hypothesise that anger needs to exceed a certain intensity threshold before rhythm is affected. If, for example, anger affects both intensity and rhythm, it is not necessarily the case that both parameters of the utterance are equally affected simultaneously or to the same extent. This cannot be a global relationship between intensity and rhythm in prosody; it may well be that the relationship is different for different emotions. So, for example, happiness may change rhythm to a kind of excited delivery before it affects intensity.

Another problem to consider is whether some of the parameters are simple effects at the moment of instantiation, or whether they are best described at the more abstract general structural description level. Are there effects which are properties only of performance and would be inappropriately characterised in any competence (static general structural description) level on the static plane? In which case, a generalisation for the potential must be stated on the static plane.

Language reflects the different gradations of types of emotive speech that a speech model needs to account for. Listeners are sensitive to quite small gradations of change and glosses on basic emotions. For example, the following language labels, among many others, are used by speakers of English with respect to reporting about themselves or noticing emotive content in others:

- *consumed by anger*–anger, global
- *twitchy with fear*–anger, local or of low intensity
- *screaming with laughter*–happiness, global and of high intensity
- *down in the dumps*–sadness, global and of low to medium intensity.

A synthesis system might perhaps best reflect these differing levels of intensity by stating a range of values on attributes in a list, one of which would be called on by a procedure when instantiated.

37

Considerations in Characterising Expression/Emotion

In order to characterise features of underlying expression, we suggest the model would ideally specify certain factors.

1 It would specify the set of vectors as distinct types of expression–for example, anger, fear, contentment or sadness.
2 It would specify an attribute on all expression/emotion types, such as *intensity*–that is, a range of the degree of the emotion type possible. For example, for *anger* the intensity could refer to the range of anger speakers might label as 'bothered, irritated, angry, furious rage'.
3 It would specify a value on intensity–for example, for 'irritated' interpreted by words such as mildly, somewhat or greatly.

We have outlined a model in Tatham and Morton (2004, pp. 322–326). The model can focus on composite expressive content or individual parameters, co-occurrence of which at particular values can be a defining factor in perceptual interpretation of what expressive type is being heard. Whichever way we look at things we have to have a formal characterisation of type of expression and its intensity, together with how it is unfolding in time (getting more/less intense etc.). The obvious reason for this is to enable the synthesis stages of a dialogue system to respond appropriately to detected expressive changes in the human user.

37.1 Suggested Characterisation of Features of Expressive/Emotive Content

The following is an XML sketch of *angry* and *happy*, two experiences of reported emotion which might commonly be encoded in the speech waveform. The assumption is that emotion can affect the physiological stance of the speaker and thus will affect the tension of the articulators, and result in some variability in the speech waveform.

37.1.1 Categories

Consider four categories to describe and relate:

Developments in Speech Synthesis   Mark Tatham and Katherine Morton
© 2005 John Wiley & Sons, Ltd.   ISBN: 0-470-85538-X

- the label on the precipitating event–e.g. ANGRY or HAPPY
- the physical correlates–e.g. fundamental frequency increase
- the cognitive correlates–e.g. intonation change
- the linguistic description–for example, e.g. intonational phrase, accent group, syllable, segment; the rhythmic unit and its constituents.

Four basic biological emotion circuits are implied which can result in four broadly specified and separable types of emotion. We encode these as *elements*. Associating a descriptive emotion label and the acoustic representation in the waveform might look like this in XML for two of these, anger and happy:

```
<emotion>
        <anger
                f0_range="+100-150"      [range is increased by
                                          100–250Hz]
                amplitude_range="+5"     [range is widened by 5dB]
                rhythmic_unit_range=     [the range is shortened by
                "-50-100">               50–100ms]
        </anger>
        <happy
                f0_range="+100-200"      [range is increased by
                                          100–200Hz]
                amplitude_range="+5"     [range is widened by 5dB]
                hythmic_unit_range=      [range is lengthened by
                "+50-100">               50–100ms]
        </happy>
</emotion>
```

In this illustration, attributes are qualified in terms of their range of values. Procedures assign the actual numbers. Attributes will vary, depending on which node in the linguistic description they can apply. For example, fundamental frequency change will occur over a longer duration, in most cases, than just on a syllable. We need to determine the amount of activity in the biological 'circuit' (Panksepp 1998) specified (e.g. its intensity), and identify combinations of circuits, for subtlety of effect, that are described as 'blends' or 'secondary emotions'–hypotheses varying the 'saturations' of what are sometimes called 'basic emotions' (Plutchik 1994). These blends can be expressed within language either by specific words or modifiers. For example, the word 'agitated' might be thought of in this model as a combination of the categories anger and fear, or the phrase *a bit jumpy* as another way of expressing a similar emotional state in a different style.

Detailed information is still not available for assigning real numbers to value ranges. We are suggesting here an approach which will characterise the type of information needed, a data structure for this information stated in XML, and a possible way to map from one type of event to another using the same general data structure. Pending more empirical evidence based on experiments to determine the actual values people might use under these conditions, the synthesis researcher will have to adopt an informed and reasoned guess as to what values to use to get the required percept of expression in the listener.

37.1.2 Choices in Dialogue Design

There is a trade-off between word selection at the cognitive linguistic level and providing the means for encoding the additional information identified as expressive content we convey when speaking. It is the case that English, at any rate, allows the subtleties of expression to be encoded either by choice of words or by the addition of expressive content to the soundwave. We assume that the words reflect the experience of subtle emotion accompanied by a cognitive intention to convey these subtleties. The message can be conveyed by words, with a minimum of emotive effect encoded in the waveform. Thus someone may say *I am angry*; but a listener might say back *You sound furious*–and the speaker might then reply *No, not furious, but really very annoyed and I didn't want this to happen.*

If some of the content is conveyed by non-verbal means, we have to ask how the signal is different from what was expected.

1 What is in the overall signal that is different?
2 Which particular features have altered?
3 What weightings on which features have changed?

The synthesis side of a dialogue system will need to know how to specify, perhaps using the vector intersection model we have described, subtle shifts from one emotion to another, the varying intensity of expression, and what might be described as 'secondary emotions', possibly characterised as combinations of basic emotion. For example, in a phone-based dialogue we might find:

Call centre synthesiser: *Please hold. One of our agents will be with you when they are free.* [neutral tone]

Human user (greatly annoyed): [mutterings to self] [detectable irritation]

Synthesiser: *Please hold. One of our agents will be with you soon.* [still using neutral tone–now inappropriate]

Human user (getting more irritated): [silence] [the silence is meaningful]

Call centre human (sing-song happy voice): *Hello, I'm Samantha. How may I help you?* [quite inappropriate tone]

Human user (resenting Samantha's tone and now angry): *I've been waiting ten minutes.* [understandable, but inappropriate tone]

The words of the synthetic voice are inappropriate, or the tone of the voice is inappropriate. Had the synthesiser attempted to anticipate the human caller's reactions to the wait etc., then Samantha's equally inappropriate tone could have improved the tension developing between the caller and the system (including Samantha). This example serves to underline the importance of adjusting expressive content in response to feedback: human beings do this on a continuous basis during a conversation. In the example the human caller is issuing all the right elements of feedback, but the call centre system is completely insensitive to these signals. Samantha has been well trained, so she will suppress her next line: *Don't you take that tone with me. I'm just doing my job.* In terms of expression a fair response to the caller, but in terms of the *flow of expressive content* appropriate for this dialogue this reply would make matters worse.

Expressive content flows in a dialogue. Dialogue models are currently very sensitive to semantic content conveyed by lexical choice (*I want to order a book*), could do more to be sensitive to pragmatic content conveyed by lexical choice (*I am irritated with your system*), and are almost always completely insensitive to expressive content conveyed by prosody ([fed up with waiting so long, so using irritated tone] *I want to order a book*).

37.2 Extent of Underlying Expressive Modelling

We have seen that generating and conveying emotion and expression are very complex behaviours for human beings. The models in psychology and linguistics are equally complex and not entirely agreed. It follows that if we feel synthesis needs to be pushed forward in this area to achieve something of the naturalness we know stems from this behaviour, we have to have some idea of how deep into the generation of the experience of human emotion do we need to go. We should ask what basic and essential points need to be taken from underlying models of language and emotion. Selecting the basics needs to be principled, and must involve characterising language and emotion and their parameterisation in such a way that we can deal with them in transparent computational terms, since the purpose is to produce computer-based synthetic voice output.

An example is the parameterisation of emotion into features, such as *intensity*, enabling us to recognise differences between distinct emotions. Intensity can be accommodated in XML code; it is, however, essential to identify the range of values that can be assigned to this attribute and the granularity of the representation. For example, do human speakers produce a meaningful continuously variable range of intensity, or is the range categorised into meaningful zones? And, in any case, is the perception of intensity focussed on continuousness or itself categorised into zones? The body of evidence on both counts points toward zoning of content along vectors, if only because categorising is a common human property. There are worries in psychology and linguistics about the detail of categorisation of continuous signals, but all are agreed on the basic idea of the need to identify zones along an otherwise continuous vector. The obvious example is the perception of a finite number of colours (an abstraction) within the continuously varying wavelength (a physical property) of light in a rainbow.

Identifying zones in expression enables us to scale down the problem. So, for example, it may be necessary to deal only with three levels of intensity simply because three levels might be the maximum human speakers recognise in speech. The practical problem, which will need empirical support from psychology experiments, is how to perform the categorisation.

The problem has already been faced in automatic speech recognition in terms of the identification of individual segments, rather than expressive content. So, for example, take the phenomenon of voice onset time (Lisker and Abramson 1964), a term in phonetics for referring to the delay in the onset of vocal cord vibration for a vowel when immediately following an initial plosive, as in the word *peep* in English. Here the delay is around 25ms, as opposed to a delay of around 5–10ms in the word *beep*. Variability enables the *p*-associated delay to range from, say, 15 to 35ms, and the *b*-associated delay to range from, say, 0 to 20ms; but the distribution of data gathered from many speakers of English will show two peaks corresponding to two perceptual categories–identified by speakers and listeners alike as the correlating physical exponents of the abstract plans for /p/ and /b/ respectively. A similar experiment conducted with French speakers in corresponding words beginning *pi* . . . and *bi* . . . (where the French vowel is not unlike the English one) might reveal a deal averaging only 5–10ms for the *pi* . . . words but −25 to 0ms for the *bi* . . . words (here the delay is actually negative, with the vocal cord vibration usually beginning *before* the release of the stop). Both languages operate or function using a pair of zones on the vocal cord onset timing vector, but the zones are located differently.

There is every reason to hypothesise that the situation is very similar with expressive content as conveyed by varying prosodic features. It is for this reason that we have favoured a combination of vectors and parameters for characterising expression. We feel that this kind of approach captures something of what is probably going on in human speakers and listeners–but, importantly for our purposes in this book, makes for a system which moves easily into the domain of synthesis. The basis for continuous variability is there, but at the same time identification and delineation of zone is in principle easy and transparent.

37.3 Pragmatics

We have concentrated on the idea that speech expression is both biologically and cognitively sourced. One source of cognitively determined expression is clearly the choice of words which convey a large part of the message; word and sentence meaning is studied by the linguistic component *semantics*. However, we suggest that another major contribution to cognitive expression is based on the *context* within which speakers and listeners find themselves. The area of linguistics that models language context is called *pragmatics*. Kearns (2000, p. 1) differentiates semantics and pragmatics:

> Semantics deals with the literal meaning of words and the meaning of the way they are combined . . . Pragmatics deals with all the ways in which literal meaning must be refined, enriched, or extended to arrive at an understanding of what a speaker meant in uttering a particular expression.

For example, a simple utterance such as *The grass is green* may presuppose this is a good thing–that is, *The grass was brown last month, but it is now green and we don't have to water it*. Or it may mean that it is now green and growing and needs cutting. It may mean the speaker is pleased, having been in the desert for a year and thus overwhelmed by the colour and amount of grass in a temperate climate.

Pragmatics is the study of information which is not directly coded in the soundwave and which represents ideas about the context used by speakers and listeners to interpret the speech soundwave. In addition to suggesting that prosody wraps the utterance, and expression wraps prosody, we suggest that pragmatics partially wraps expression (Morton 1992). Other 'channels' contributing to expression are described by *semantics*, and cognitive constructs of emotion such as appraisal. Thus, expressed formally the relationship might look like

```
<pragmatics>
        <expression>
                <prosody/>
        </expression>
</pragmatics>
```

Many approaches to pragmatics have been proposed and are discussed by Levinson (1985) and Verschueren (2003). For illustrative purposes we paraphrase Grundy (2000) about some of the general types of pragmatic effects that can be characterised by categories relating to the utterance, such as these:

- *Appropriateness*–utterances must fit the context. *The sun is shining* is not an expected remark at midnight in England, but will be acceptable for California where it is still only 4 pm. *Please hold the line* is appropriate if the delay will be short and the line need not be held for very long.
- *Indirect meaning*–the literal, or semantic, meaning is not what is intended. For example, the greeting *How are you?* does not mean that you necessarily want to know about the listener's recent accident. Or, for example, *Please hold the line* is a request to the caller to continue listening for a reply to a query, but this may be confusing to someone who is not familiar with the phrase *hold the line* when using a phone.
- *Inference*–an action by the listener of adding information to the literal meaning, based on knowledge of the world. For example, your neighbour says *I'd like a word with you*. You may well become apprehensive that you about to hear a complaint about your barking dog. Similarly, *Please hold the line* is based on the expectation that a person will soon answer your query and perhaps discuss your problem.

These examples of types of pragmatic effect have in common that the surrounding language context can be described by sentences that are not linguistically similar to the actual utterance. That is, pragmatics is not a paraphrase but describes possible 'missing' linguistic structures which might add information to the utterance itself. It is as if these sentences are required to exist in some form–if only an 'understood' form–before the utterance is decoded. These are sentences that could be spoken leading up to the utterance, except that we do not have the time to form them, or even perhaps *think* them before speaking. Pragmatics focusses on actual utterances, and characterises what the speaker intends with varying degrees of complexity.

In computer speech, it is also useful to determine what kind of voice is best for the application, bearing in mind the type of listener–for example, their age range, gender, culture, education, ability to speak the language, and so forth. These characteristics have some role in pragmatics by narrowing down the field of inference. For example, inferences made if a child is speaking may be different from those made from adult speech. Some understanding of the behaviour of the listener upon receiving the message might be useful. Inferences made by the listener might elicit unwanted behaviour. For example, on the phone, will the listener

- hang up if delayed?
- be too nervous to reply to a non-human computer voice (not'respond to a too decisive voice)?
- be abusive if irritated by repetition?
- be incoherent if distressed?
- be unable to understand the language?

How much the listener feels threatened (fear) and how much he or she feels able to cope (anger) will guide that person's future interaction with the system.

One implication for synthesis is that even if computer speech is used within a specified context, such as giving information or instructions from an emergency call centre, then a standard phrase like *Please hold the line* conveying a particular attitude might be very important for the best use of the system. And the very action of generating computer speech across cultures itself presupposes that attitudes are universal–that a speech waveform conveying a decisive attitude will be decoded as decisive and appropriate for an emergency to *all* listeners. There might be a presupposition that listeners will interpret decisive as insulting, laconic, etc.

In our example utterance *Please hold the line* the following pragmatic effects might contribute to its expressive content:

- *presupposition*–the listener requires advice from this phone number
- *implication*–the listener is actively looking for advice
- *situational*–the listener should not be alarmed by an anxious computer voice.

Summing the pragmatic considerations might well suggest that in most circumstances the best voice would be a decisive voice for most callers. This overall goal, decisiveness, wraps the utterance. However, changing the pragmatic conditions would suggest another type of voice: *Please hold the line* spoken for the third time could well be irritating. The condition 'situational' might dominate and suggest a friendly soothing tone would be better. The listener's behaviour needs to be predicted by the system in the hope that most listeners will be able to behave efficiently and appropriately for the situation. There seems to be a rather blurred area between *situational* and *social context* which perhaps need more fine-tuning (Johnson *et al.* 2004; Van Lancker-Sidtis and Rallon 2004).

It is also not clear what relation expressive/emotive content might be posited with respect to a pragmatic modelling in speech. Levinson (1985, p. 41) refers to 'emotive function on the speaker's state' as one of the fundamental constituents of communication. He discusses earlier work by Jacobson (1960). Verschueren (2003) notes that some classes of verbs imply emotive content–for example, 'assertive (expressing a belief), all directives (conveying a wish) . . .'. He lists not only verbs but several dozen sentence types that might fall into an emotive category–for example, warning, advising, praising, greeting. Specifying a relation between a narrow definition of expression, emotion and pragmatics seems to be a question for researchers in the future. Even so, in general however, we think that introducing pragmatics–about inferences etc. made by the listener depending on the interpretation of the waveform–could well contribute to building more successful speech communication systems.

38

Summary

The model we suggest integrates two modelling features:

- expression as overall wrapper, and
- the notion of opposing static and dynamic characterisations.

An individual wishes to express a thought or emotion, or both together, to a another individual. The overall expression can be conveyed through various modalities, including writing, dancing, speaking etc.

38.1 Speaking

If speaking is chosen, a sequence of events unfolds.

1 The thought (the plain message) is encoded as language by accessing static knowledge bases which characterise all possibilities for each component separately (separation is a productive concept in the modelling process, but does not necessarily reflect what actually happens in the human being, of course):

- *semantics*–cognitively specified (cognitively sourced emotion can feed into this component)
- *syntax*–cognitively specified (cognitively sourced emotion can feed into this component)
- *other knowledge bases* such as cultural, social context etc.

2 The output resulting from accessing and processing this knowledge is expressive speech (or writing).
3 Speaking calls for

- accessing a *static plane* where the phonological knowledge base is found. This plane contains hierarchically specified units (like syllables and wider domains, such as accent groups and intonational phrases used in intonation), together with a full characterisation of the data structures in which they operate.
- deriving an output from the *dynamic plane* in the phonological utterance plan which specifies the sound pattern for an utterance, and implicitly contains information from syntax, semantics and such cognitively sourced expressive/emotive content that may be encapsulated in the choice of words or syntax.

Developments in Speech Synthesis Mark Tatham and Katherine Morton
© 2005 John Wiley & Sons, Ltd.   ISBN: 0-470-85538-X

- rendering the plan by moving it to the dynamic phonetic plane which accesses the phonetic static plane holding the rendering knowledge base (phonetics). This plane contains information about how to render the phonological plan. The output is characterised symbolically as a string of intrinsic allophones for this particular utterance. In the complete simulation–synthetic speech–the output is an acoustic signal embodying all properties necessary for the appropriate decoding by a listener of the plain message and expressive content intended by the speaker or biologically added by their stance.

The speaker will be in a constantly changing environment as speaker and listener interact– a behaviour to be characterised in the dialogue model. Information is needed about the changing physical state of the speaker. We could have on the static plane a statement of all possible physical states of the speaker and their consequences, but the list would be impossibly large to calculate. Some of the physical states will arise because the speaker is reacting in an emotive way to the environment. And some other types of information may be relevant, such as speech handicap, food in the mouth, pipe in the mouth, or perhaps an inability to move and/or speak in general–such as dysarthria or standing on head.

> The 'speaking device', the vocal tract, is not a rigid structure but a mutable one. It is constantly changing with tonic activity, responding to breathing, swallowing, unintended tongue movements, when not speaking; when speaking, the movements are controlled and intended, and alter the shape of the tract.

There is a working space in contact with the phonetic dynamic plane which acts upon the plane to coordinate phonetic activity, logically prior to the movement of the neuromuscular system and its motor control system. This working space contains the *cognitive phonetic agent* (the CPA) which, among other possible functions, assembles the information about the state of the physical system in general, and specifically about the vocal tract which participates as a structure in the general physical system. For example, if the individual is tense in general, the vocal tract will also be in a state of tension.

> The CPA is seen as one of a class of agents (Garland and Alterman 2004) organising and monitoring the individual. A similar agent might take the basic process of walking and turn it into the much more sophisticated and carefully controlled activity of dancing.

The CPA on the phonetic dynamic working space knows about the plasticity of the physical system, its limits and constraints (because the physical condition is measurable and real). The CPA is the device which does the work of conveying the thoughts and emotion to the outside world by controlling the movement of the articulators through fine-tuning of the neuro-muscular system (Fowler 1980). So the CPA performs a monitoring and coordinating role, and in particular supervises and oversees the speaking operation (Morton and Tatham 2003).

38.2 Mutability

All phonetics models treat the vocal tract as having a fixed specification within one speaker. Only in a trivial way is allusion made to mutability as when phoneticians refer to temporary changes in, say, voice quality when someone has a cold. But the reason why we have to stress this point here is that mutability is regarded by listeners as part of the distinguishing characterisation of what constitutes a *human* voice. We could exaggerate a little by saying that it is factors like this, together with varying expressive content, which *are* the naturalness of human speech. For historical reasons phonetics does not make such issues central to the discipline, and linguistics as a whole has virtually nothing to say about anything other than the basic linguistic functions of language.

The output of our dynamic phonetic working surface is the input to the neuromuscular system which handles control of the vocal tract. But there is a jump here in terms of representation. The representation of the phonetic output may need a process of re-representation to make it suitable as the input to the neuromuscular processes (Tatham and Morton 2004). One further problem is that we know that it is possible to interfere with neuromuscular processes on a short-term *ad hoc* basis. This would involve a direct line from inside the dynamic phonetics to inside the neuromuscular processing–perhaps not *via* the output/input levels of representation.

The speech waveform encodes aspects of language characterised in linguistics: semantics, pragmatics, syntax, phonology. In addition it encodes cognitively sourced emotive effects, physical properties of the system, biologically sourced emotive effects. There are other effects as well, but these are the main sources having an influence on the final waveform. These influences are ultimately reflected in the acoustic signal and the major parameters we recognised in our acoustic models, features such as

- fundamental frequency, including its range and dynamic movement within that range
- formant frequencies, amplitudes and bandwidths, including ranges and dynamic movement
- the timing of utterances, reflected in the overall rhythm of the utterance, but including local variations in the timing of rhythmic units and their included units–syllable and phone-sized segment.

We must not forget asynchronous changes in the articulatory and acoustic parameters of a speaker which researchers like Firth (in an abstract characterisation) and Browman and Goldstein (in a more physical way) attempted to capture. Figure 38.1 shows a diagram of the model.

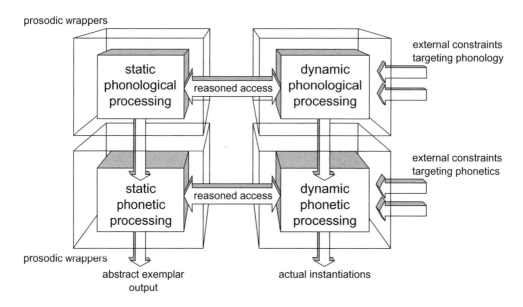

Figure 38.1 Simplified diagram of the overall model showing the relationship between static processes and dynamic processes. Note that only an abstract exemplar output is permitted on the static plane: actual instantiations of plans are output only from dynamic processing.

Part X

Concluding Overview

Shared Characteristics Between Database and Output: the Integrity of the Synthesised Utterance

In long-unit or variable-unit waveform concatenation systems (i.e. those which go beyond simple diphones as their units) it is important to remember that, although the output corresponds to the utterance plan, which in turn is dependent on the originating sentence or concept, there are nevertheless characteristics which the output inherits from the database. A clear understanding of these characteristics and what they mean to the *integrity* of the synthesised speech is important. So, for example, features carried over include

1 Voice characteristics (phonetic):

- gender
- identifying properties of the original voice.

The voice which made the recordings usually has a well-defined and recognisable gender. It also has identifying properties, properties which enable a listener to name the speaker and which, while not individually unique to this speaker, in combination are likely to be so. Speakers are usually unable to alter these properties without a great deal of effort, and even then it is questionable whether they can consistently 'act' a different voice.

2 Style and accent (phonological and phonetic):

- general speaking style
- regional or social accent (not dialect, which is predominantly lexical and syntactic rather than phonological and phonetic).

The recorded database will have a general speaking style which is also a property of the original voice, and it will also include any regional or social accents. Some researchers believe that it is possible to put together a 'neutral' database, but this is theoretically impossible (Tatham and Morton 2004) even when using trained actors as recording subjects.

3 Expressive content (phonological and phonetic):

- general tone of voice
- emotional and other pragmatically sourced properties.

At the time of the recording the speaker will adopt a general tone of voice which will include general expressive content. In order to read aloud the speaker must understand the text; he or she will not be able to conceal reaction to the semantic and pragmatic content of the text itself which may reveal itself to the listener even when the database is reused for novel utterances during the later synthesis phase.

4 General rendering effects:

- individual coarticulatory properties
- cognitive phonetic preferences.

Developments in Speech Synthesis   Mark Tatham and Katherine Morton
© 2005 John Wiley & Sons, Ltd.   ISBN: 0-470-85538-X

If we adopt a general speech production model which includes the concept of concatenated–but otherwise isolable–sound segments, we also have to include the concept of coarticulation, characterised within the *theory of coarticulation*. Coarticulatory effects are dependent on the physical characteristics of speech (the mechanics and aerodynamics) and are always time-governed. Although there are generalisable properties of coarticulation there are undoubtedly personal or idiosyncratic properties, since each individual is different with respect to details of their anatomy etc. In addition, there may be accent or idiosyncratic manipulation of coarticulatory effects–there are characterised within the *theory of cognitive phonetics*. These effects are likely to be revealed to a listener, even when the recording is dissected and reassembled.

5 Meaning-specific rendering effects:

- 'interpretive variations' in precision.

Despite the effects characterised in general by the theory of coarticulation, it is nevertheless the case that speakers have a certain amount of control over the way segments concatenate. Under (1) above we mention general preferences by the speaker for the way coarticulation is controlled, but there is also a degree of *local* variation in utterance precision. This is important because it derives from the speaker's understanding of what he or she is saying–its meaning. But the original meaning of the speech (in the database) is not the same as the later meaning of utterances to be synthesised. Although we can theorise about how we might alter these local precision effects, no provision has yet been made in concatenative systems for including such effects. However, it is clear that listeners are sensitive to precision (it reveals some of the speaker's personality and attitude to what is being said and to who is being spoken to) and can probably easily detect whether the new utterance is adequately natural in this respect. Too much precision here, or too little there, will be detected.

It can be argued that many of these properties are not properties of individual segments but are associated more with longer term, supra-segmental or prosodic features. While this is true from our normal analytical perspective–that is, for example, intonation is always characterised as a long-term phenomenon–these features have segment-by-segment properties also. These properties will be preserved, if only on a statistical basis. If, for example, the speaker has a generally high-pitched voice, a description based on a long-term analysis rather than on the analysis of the fundamental frequency of an individual segment, there will be properties associated with a high pitch throughout the database. These may include higher than average formant frequency values or a higher than average percentage of aperiodic sound in the voice source. In more general terms there are basically two types of variable of this kind in speech:

1 generalised long-term global variables characteristic of the overall properties of the speaker
2 specialised short-term local variables reflecting the way the speaker is responding to specific semantic or pragmatic properties of what is being said.

In synthetic speech (1) is acceptable, if not desirable, because it guarantees the internal integrity of novel utterances synthesised from the database. However, (2) presents a serious challenge for synthesis because the effects here are directly derived from the properties of the *original* text (or conversation) recorded–and the novel utterances are guaranteed to be different from this. For this reason, (2) counts *against* internal integrity of novel utterances.

Future generations of synthesiser will put full weight on the concept of integrity of the speaker. Up to the present time we have had semantic integrity: what the synthesiser says is meaningfully relevant. And in dialogue systems the dialogue control procedures try to ensure integrity of pragmatic content, at least as far as lexical selection or choice of words is concerned. However, it is becoming clear that there are local as well as global considerations at issue for integrity.

The concept of phonological and phonetic integrity has received little attention; the hope has been that the long-term or global characteristics referred to above will take care of this. One reason for this neglect is that phonetic and phonological analysis leads usually to a symbolic representation which is, by its nature, insensitive to the detail of integrity. This leads in turn to a markup of the database which itself neglects these characteristics because it depends by its very nature on normalising out the detail which often leads to integrity–at best a feed-forward procedure which worsens the effect.

Concept-To-Speech

Hirose and Minematsu (2002) describe a concept-to-speech synthesis system for use in a dialogue environment. This is true concept-to-speech synthesis since replies in the dialogue situation can be generated without the need to introduce an orthographic representation of the reply utterance. Text is explicitly avoided because at its simplest–that is, with no special markup–it fails to encode much phonological/prosodic or phonetic information: the content of text reflects syntactic and semantic, lexical and morphemic content.

An important question for any concept-to-speech system is whether it might be appropriate to have an intermediate stage between 'pure concept' and phonological representation, and what form it would take. For example, a text system augmented by XML markup of the missing information would possibly fit the bill. Here it would a question of identifying what properties of the concept representation are needed for improved synthesis but which are not able to be encoded by a plain text system. These properties could be explicitly added to plain text with XML-style markup.

Even in dialogue systems, though, it is easy to imagine that text input will be necessary, as when the dialogue is about some written material. When text *is* involved, whatever the reason, there arises the question as to whether the text itself should be allowed to continue to dominate synthesis strategies. Text orthography, for all its idiosyncrasies, does encode most segmental aspects of speech, even when it is never intended to be read out loud. But suppose we take the line that, instead of allowing text to dominate in all text-based situations, we allow prosody to dominate. Prosody is much more closely related to concept in many ways. For example, the concepts <query> or <question> can link directly to a question intonation contour in phonological prosody, and eventually to a question fundamental frequency contour in phonetic prosody. Concepts like <uncertainty> and <authority> also link more directly to prosody than they do to any general properties of text (excluding deliberate lexical selection). We have not undertaken in this book to discuss the representation of concept–the area is far too new and unstable an idea at the moment, and research is really only just beginning. But we can see that certain aspects of an interim solution could be designed to facilitate a future move to, or experimentation in, concept based synthesis.

Pragmatic considerations we discussed in Chapter 37 link concept on the one hand and expressive prosody on the other, and it is here that we can experience the real beginning of how we might handle concept in true concept-based dialogue systems. Already dialogue systems involve recognisable 'dialogue management' evident in a sub-system responsible for directing, in effect, the semantic flow of the dialogue. Our model of human speech has suggested that for speech simulation in synthesis phonological choice and phonetic variability be under the management of the cognitive phonetics agent. Many variability considerations which are governed by pragmatic considerations, such as much of the expressive content of speech, derive ultimately from considerations of concept and concept flow. Indeed, pragmatic processes will have a direct input from concept management.

This brief discussion of concept-to-speech synthesis is intended to make a number of points.

1 Concept-to-speech synthesis need not imply an extreme abandonment of text, since 'real' text undoubtedly occurs anyway from time to time in concept-based systems.
2 The advantages of concept-based synthesis in, say, dialogue lie in the fact that it is not dominated by text.
3 Semantic and pragmatic considerations determine how the dialogue flows, is tracked and managed.
4 Many if not all of the non-lexical aspects of dialogue content and management become encoded eventually as prosody.
5 Prosody in concept-based systems can 'evolve' prior to the emergence of lexical content.

Text-To-Speech Synthesis: the Basic Overall Concept

Elsewhere we discuss the basic components of text-to-speech synthesis systems. These include

- normalisation of the incoming text;
- pronunciation derivation by means of a large dictionary, accompanied by
- sets of rules or some other method (e.g. the use of an artificial neural network) for converting the orthographic representation to an underlying phonological representation;
- semantic and grammatical parsing, to assist
- the prediction of an appropriate prosody to fit to the phonological string.

The output of these components is a symbolic representation of what is to be spoken, including its segments and prosodic contours. There is considerable confusion over terminology, but essentially the output is usually an extrinsic allophonic string marked with respect to an equivalent level of prosodic information.

The level of abstraction for this representation, which corresponds to our *utterance plan*, is important. Extrinsic allophones are symbols which represent segments derived from their underlying phonemic characterisations and which include the results of phonological processes. Phonological variation is cognitively determined and must be distinguished from phonetic variation–usually called coarticulation–which is essentially governed physically rather than cognitively. The equivalent prosodic information marked on the segmental string is derived by phonological prosodic processes, once again to be distinguished from phonetic prosodic processes which are analogous to coarticulation of segments.

There are various techniques for combining prosodic and segmental information, but one way is to assume the dominance of the segmental information and express it as a string of elements which possess prosodic attributes. We prefer to assume the dominance of prosodic information and assign segmental attributes. Either way, this representation is the one which is to be rendered in the low-level part of the text-to-speech system. Notice that our account of the typical system concludes the high-level part with an abstract symbolic representation. We adopted this approach from human speech production models which make a sharp distinction between phonological (high-level cognitive) and phonetic (low-level physical) processes. Speech production models of this type conform explicitly to contemporary ideas in linguistics and psychology.

We believe it is a potentially confusing conceptual mistake to mix abstract and physical attributes at this stage. Thus systems which mark a symbolic representation of segments with physical quantities such as duration or fundamental frequency violate an important principle–the avoidance of mixing abstract and concrete characterisations. This is not an idiosyncratic belief. For example, on a practical level, by making the distinction rigorously in SPRUCE we were readily able to test the high-level components against a variety of different low-level systems with the minimum of adjustment to the interface. On a theoretical level it makes sense to speak of an abstract prosodic contour so that different rendered versions of the single contour can be compared as between different human speakers saying the 'same' utterance. It is theoretically important in linguistics and psychology to capture the human notion of *sameness*, and we believe it is important to transfer this notion to synthesis systems. Sameness resides at the well-defined symbolic output from the high-level system, whereas detailed physical variability comes in later.

Prosody in Text-To-Speech Systems

Most text-to-speech systems introduce breaks into the input word stream, and these occur at what are often referred to as 'major syntactic boundaries'–for example, between sentences, immediately before relative pronouns, etc. This syntactic parse of often quite rudimentary, its main purpose being to establish prosodic breaks in utterances; the full-scale parse of the computational linguists is rarely called for. Once breaks delimiting mostly intonational phrases are established, symbolic prosodic markup can proceed. Most systems provide a means for marking stressed and unstressed syllables (thereby establishing rhythmic groups or accent groups), with these superimposed on generalised contours. In the simplest case–for example, a short sentence consisting of a single intonational phrase–a contour would be a relatively linear intonational rise to the focus of the phrase (say, the main word) with a fall from this point to the end of the phrase. All these markers or tags come from phonological prosody and are abstract; they bare only a notional relationship with the physical parameters available to phonetic prosody. Below is an example of a simple utterance markup in this way using intonational assignment rules of SPRUCE (see Chapter 32 for details of the markers and the assignment procedure). This illustration is adapted from Morton *et al.* (1999). The phonetic symbols represent the segmental utterance plan, and are typical for Southern British English.

Planned utterance (in orthography):

Capital initials can, if the typography allows it, be rendered by small capitals.

Phonological intonation wrapper markup (in SPRUCE notation):

	kæ	pɪt	əl	ɪ	nɪ	ʃəlz	kæn				
H[	S	U	U	U	S	U	U	T–	]H		

	ɪf	ðə	taɪ	pɔ	grəf	i	ə	laʊz	ɪt		
L[	U	U	U	S	U	U	U	S	S	T–	]H

	bi	ren	dɜd	baɪ	smɔl	kæ	pɪt	əlz		
L[	S	S	U	U	S	F	U	U	T–	]L

It is important not to confuse this markup with a representation of fundamental frequency. The two are related in a fairly complex nonlinear way, and efforts to characterise the relationship still fall short of providing an entirely natural sounding output. Part of the problem is the abstract phonological markup and part of the problem is how this is converted into the available physical parameters. We can mention, for example, the notion of *declination*. Researchers have observed that in most languages during short plain statements the fundamental frequency drifts downward throughout the utterance. We like to think of the phenomenon as intrinsic to the way the aerodynamic system works, reflecting a tendency for sub-glottal air pressure to fall continuously between breaths. Many researchers disagree with this model, citing examples of rising fundamental frequency, say, during some types of question in many languages; they also allude to some stylistic and expressive effects which produce a rising fundamental frequency, or at least one which does not fall continuously in the way the 'intrinsic phenomenon' model would predict. Our answer to this interpretation is simple, and is the same as our answer for unexpected coarticulatory effects between segments.

1 Intrinsic phenomena are physical, nothing to do with speech, and their existence cannot be denied. The articulators have inertia, speaking progressively reduces sub-glottal air pressure etc.
2 *But* this does not mean that we cannot interfere with many of these processes. They may be intrinsic and physical, but they are able to respond to cognitively originated control–at will many of them can be partially negated or even enhanced.

A combination of these two considerations is enough to satisfy all of the data so far observed in connection with variable coarticulation and intrinsic aerodynamic phenomena, such as declination and vocal cord vibration failure. Manipulation of effects such as these, however, probably entails careful supervision–hence our repeated reference to the CPA (cognitive phonetics agent) whose job it is to manage rendering in general, especially working out how to cope with intrinsic phenomena of physical origin which may have a serious effect on the 'perceivability' of the output waveform.

Optimising the Acoustic Signal for Perception

In speech synthesis terms, the equivalent to the cognitive phonetics agent in our model of human speech production is responsible for optimising the signal for perception. Optimisation does *not* mean making sure segmental targets are hit, or somehow making sure that the acoustic signal reflects careful or precise articulation. Human speech is not like this. Optimisation should be interpreted only as providing the most *appropriate* signal. Here are just a few examples.

- If a word is likely to be confused with some other word, it should be pronounced more carefully or more slowly than words that are unlikely to be confused.
- If it is stylistically important to emphasise a coarticulatory effect, then do so–as, for example, among young people in certain American English accents when *Man!* is spoken with emphasis (the degree of inter-nasal nasalisation of the phonologically oral vowel /æ/ is *deliberately* enhanced, though it must *always* be present to some degree in *all* accents).
- If the conversation is relatively intimate, there can be a general relaxation of overall precision of articulation.
- If the speaker is irritated or scolding, despite a general increase in rate of delivery, coarticulation is *reduced*–the opposite of what a general coarticulation model would predict because the phenomenon's intensity usually correlates more or less with rate of delivery: the faster the speech the greater will be the effects of coarticulation.

These few examples are all basically about segment realisation. But phonetic prosody tackles very similar longer term or suprasegmental effects. We have given the example of *controlled* declination, but we could equally well cite the wide fluctuations in local fundamental frequency associated with surprise on the part of the speaker, or the changes made to overall amplitude to control emphatic stressing, etc.

We have digressed a little into this area of the subtle cognitively dominated control of intrinsic phenomena of physical origin because the notion is central to the idea that how we make sure that speech–human or synthetic–is appropriately optimised for perception. The reason why it needs to be optimised is quite simply that this is what listeners expect: it is part of the humanness of speech, and therefore something we feel is a major contributor to judgements of naturalness. Re-orienting our synthesis models as incorporating predictors of perception is one of the next major tasks for synthesis, and we cannot escape the impact it will have on present strategies. Changes in the way we approach the high-level system will be relatively easy and might simply involve shifting dominance to prosody rather than segmental aspects of speech. But some of the biggest and most difficult changes will come at the low-level area where the actual acoustic signal is produced. Imagine, for example, having to index segments in the database of recorded material in a unit selection system with how susceptible they are to de- or re-coarticulation. Or imagine the enormity of the signal processing task which has to deal with the acoustic results of *continuously variable* precision of articulation–for that is exactly how it is in real speech.

The continuous variability of precision we observe in real speech in, say, a conversational situation is a crucial factor in estimates of naturalness. The variability correlates with linguistic, psychological, social, expressive, acoustic and other environmental features, some of which are embedded in the utterance (e.g. phonological contrast with another more predictable word), while others are external (e.g. background noise). The following simplified diagram shows the main components of the model for both human speakers and text-to-speech synthesisers.

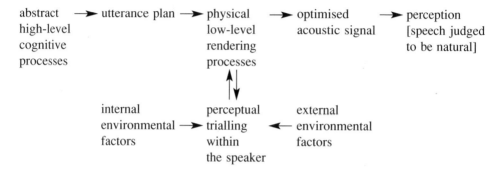

In the diagram we see that high-level cognitive processes provide an utterance plan for low-level rendering. Part of the rendering process involves perceptually trialling a hypothesised acoustic signal: this is done within the human speaker's mind. Adjustments are made to rendering in an iterative fashion so that the final acoustic signal can be optimised for perception. The speaker, by pre-trialling the 'sound' has ensured that it is appropriate for perception. Or, the synthesiser has ensured that the signal will be judged natural (at least from this point of view).

Conclusion

There have been a number of important advances made in speech synthesis during the past few years, not least in the area of prosody. What is emerging, though, is that we still have a long way to go before synthetic speech, particularly when used in an interactive environment, becomes completely natural and acceptable to lay users. We are beginning to see that one reason for this is the fact that researchers have underestimated the shortfall in our understanding of how speech is produced and used by human beings. In general we had underestimated the role of variability and how it relates to the perception of the signal; and in particular our models of prosody and expressive content have proved inadequate for the needs of producing truly convincing synthesis.

We end with two quotations from the Keynote Address given by Gunnar Fant to the Tenth International Congress of Phonetic Sciences which took place in Utrecht in August 1983 (Fant 1983):

> We now approach the general problem of speech research in quest of the speech code and the relation between message units and their phonetic realization with all the variability induced by contextual factors including language, dialect, speaker specific situational and stylistic variations. It would be a wishful dream to extract all this knowledge merely

by computerized statistics; i.e., to collect a very large material of speech, give the computer some help for segmenting transcriptions, and look up and then just wait for the results to drop out. (p. 26)

We need an output-oriented linguistically based approach to obtain a constructive perspective. Invariance and variability have to be structured by far-reaching rules which at present rely on ad hoc, or merely qualitative insights in speech production and perception. (p. 20)

References

Adolphs, R. and Damasio, A. (2000). 'Neurobiology of emotions at a systems level', in J. Borod (ed.), *The Neuropsychology of Emotion*. Oxford: Oxford University Press.

Allen, J., Hunnicut, S. and Klatt, D. (1987). *From Text to Speech: the MITalk System*. Cambridge: Cambridge University Press.

Artstein, R. (2004). 'Focus below the word level', *Natural Language Semantics* **12**: 1–22.

Averill, J. (1994). 'In the eyes of the beholder', in P. Ekman and R. Davidson (eds), *The Nature of Emotion: Fundamental Questions*. Oxford: Oxford University Press, 7–19.

Bechtel, W. and Mundale, J. (1999). 'Multiple realizability revisited: linking cognitive and neural states', *Philosophy of Science* **66**: 175–207.

Black, A. and Campbell, N. (1995). 'Optimising selection of units from speech databases for concatenative synthesis', in *Proceedings of Eurospeech 95*, Madrid, Vol. 2: 581–584.

Black, A. and Font Llitjós, A. (2002). 'Unit selection without a phoneme set', in *Proceedings of the IEEE Workshop on Speech Synthesis*, Santa Monica: CD-ROM.

Black, A. and Taylor, P. (1994). 'Synthesizing conversational intonation from a linguistically rich input', in *Proceedings of the 2nd ESCA/IEEE Workshop in Speech Synthesis*, New-Paltz, New York: 175–178.

Borden, G., Harris, J. and Raphael, L. (1994). *Speech Science Primer: Physiology, Acoustics, and Perception of Speech*. Baltimore: Williams & Wilkins.

Borod, J. (1993). 'Cerebral mechanisms underlying facial, prosodic and lexical emotional expression', *Neuropsychology* **7**: 445–463.

Borod, J., Tabert, M., Santschi, C. and Strauss, E. (2000). 'Neuropsychological assessment of emotional processing in brain-damaged patients', in J. Borod (ed.), *The Neuropsychology of Emotion*. Oxford: Oxford University Press, 80–105.

Bregman, A. (1990). *Auditory Scene Analysis: the Perceptual Organization of Sound*. Cambridge, MA: MIT Press.

Browman, C. and Goldstein, L. (1986). 'Towards an articulatory phonology', in C. Ewan and J. Anderson (eds), *Phonology Yearbook 3*. Cambridge: Cambridge University Press, 219–253.

Bulyko, I. and Ostendorf, M. (2002). 'A bootstrapping approach to automating prosodic annotation for limited-domain synthesis', in *Proceedings of the IEEE Workshop on Speech Synthesis*, Santa Monica: CD-ROM.

Campione, E., Hirst, D. and Veronis, J. (2000). 'Automatic stylisation and modelling of French and Italian intonation', in A. Botinis (ed.), *Intonation: Research and Applications*. Dordrecht: Kluwer, 185–208.

Carlson, R. and Granstrom, B. (1976). 'A text-to-speech system based entirely on rules', *Proceedings of ICASSP 76*. Philadelphia: 686–688.

Carmichael, L. (2003). 'Intonation: categories and continua', in *Proceedings of the 19th Northwest Linguistics Conference*. Victoria, BC.

Chomsky, N. (1957). *Syntactic Structures*. The Hague: Mouton.

Chomsky, N. (1965). *Aspects of the Theory of Syntax*. Cambridge, MA: MIT Press.

Chomsky, N. and Halle, M. (1968). *The Sound Pattern of English*. New York: Harper & Row.

Clark, J. and Yallop, C. (1995). *An Introduction to Phonetics and Phonology*. Oxford: Blackwell.

Clocksin, W. and Mellish, C. (1994). *Programming in Prolog*. Berlin: Springer-Verlag.

Clore, G. and Ortony, A. (2000). 'Cognition in emotion: always, sometimes, or never?', in R. Lane and L. Nadel (eds), *Cognitive Neuroscience of Emotion*. Oxford: Oxford University Press, 24–61.

Developments in Speech Synthesis   Mark Tatham and Katherine Morton
© 2005 John Wiley & Sons, Ltd.   ISBN: 0-470-85538-X

Comrie, B. (1989). *Language Universals and Linguistic Typology: Syntax and Morphology.* Chicago: University of Chicago Press.

Cooke, M. and Ellis, D. (2001). 'The auditory organization of speech and other sources in listeners and computational models', *Speech Communication* **35**: 141–177.

Cruttenden, A. (2001). *Gimson's Pronunciation of English.* London: Arnold.

Dalgleish, T. and Power, M. (1999). 'Cognition and emotion: future directions', in T. Dalgleish and M. Power (eds), *Handbook of Cognition and Emotion.* Chichester: Wiley, 799–805.

Damasio, A. (1994). *Descartes' Error: Emotion, Reason, and the Human Brain.* New York: Penguin Putnam.

Davidson, R. (1992). 'The neuropsychology of emotion and affective style', in M. Lewis and J. Haviland (eds), *Handbook of Emotions.* New York: Guilford Press, 143–154.

Davidson, R. (1993). 'Parsing affective space: perspectives from neuropsychology and psychophysiology', *Neuropsychology* **7**: 464–475.

Descartes, R. (1649). *The Philosophical Writing of Descartes*, trans. J. Cottingham, R. Stoothoff and D. Murdoch (1984–1991). Cambridge: Cambridge University Press.

Di Cristo, A. and Hirst, D. (2002) 'De l'acoustique à la phonologie, représentations et notations de l'intonation: une application au Français', in A. Braun and H. Masthoff (eds), *Phonetics and its Applications: Festschrift for Jens-Peter Köster on the Occasion of his 60th Birthday.* Stuttgart: Franz Steiner Verlag.

Drullman, R. and Collier, R. (1991). 'On the combined use of accented and unaccented diphones in speech synthesis', *Journal of the Acoustical Society of America* **90**: 1766–1775.

Dutoit, T. (1997). *An Introduction to Text-to-Speech Systems*, Dordrecht: Kluwer.

Dutoit, T., Bataille, F., Pagel, V., Pierret, N. and van der Vreken, O. (1996). 'The MBROLA Project: towards a set of high-quality speech synthesizers free of use for non-commercial purposes', in *Proceedings of the International Congress on Spoken Language Processing*, Philadelphia.

Ekman, P. (1992). 'An argument for basic emotions', *Cognition and Emotion* **6**(3/4): 169–200.

Ekman, P. (1999). 'Basic emotions', in T. Dalgleish and M. Power (eds), *Handbook of Cognition and Emotion.* Chichester: Wiley, 45–60.

Ekman, P. and Davidson, R. (1994). 'Afterword: are there basic emotions?', in P. Ekman and R. Davidson (eds), *The Nature of Emotion: Fundamental Questions.* Oxford: Oxford University Press, 45–47.

Epstein, M. (2002). *Voice Quality and Prosody in English.* PhD thesis, University of California at Los Angeles.

Fant, G. (1983). 'Phonetics and speech technology', *Quarterly Progress and Status Report 2–3*, Stockholm: KTH, 20–35.

Firth, J. (1948). 'Sounds and prosodies', *Transactions of the Philological Society* 127–152. Reprinted in W. Jones and J. Laver (eds), *Phonetics in Linguistics: a Book of Readings.* London: Longman.

Fowler, C. (1980). 'Coarticulation and theories of extrinsic timing', *Journal of Phonetics* **8**: 113–133.

Frijda, N. (1993). 'Moods, emotion episodes, and emotions', in M. Lewis and J. Haviland (eds), *Handbook of Emotions.* New York: Guilford Press, 381–403.

Frijda, N. (2000). 'The psychologists' point of view', in M. Lewis and J. Haviland-Jones (eds), *Handbook of Emotions.* New York: Guilford Press, 59–74.

Garland, A. and Alterman, R. (2004). 'Autonomous agents that learn to better coordinate', *Autonomous Agents and Multi-Agent Systems*: **8**(3): 267–301.

Gee, J. and Grosjean, F. (1983). 'Performance structures: a psycholinguistic and linguistic appraisal', *Cognitive Psychology* **15**: 411–458.

Goldsmith, J. (1976). *Autosegmental.* PhD thesis, MIT; also New York: Garland Press.

Grabe, E., Post, B., Nolan, F. and Farrar, K. (2000). 'Pitch accent realisation in four varieties of British English', *Journal of Phonetics* **28**: 161–185.

Grundy, P. (2000). *Doing Pragmatics.* London: Arnold.

Hardcastle, W. and Hewlett, N. (1999). *Coarticulation: Theory, Data and Techniques.* Cambridge: Cambridge University Press.

Harré, R. and Parrott, W. (1996). *The Emotions: Social, Cultural and Physical Dimensions of the Emotions.* London: Sage.

Hayes, B. (1995). *Metrical Stress Theory: Principles and Case Studies.* Chicago: University of Chicago Press.

Hess, W. (1983). *Pitch Determination of Speech Signals.* Berlin: Springer-Verlag.

Hertz, S. (2002). 'Integration of rule-based formant synthesis and waveform concatenation: a hybrid approach to text-to-speech synthesis', in *Proceedings of the IEEE Workshop on Speech Synthesis*, Santa Monica: CD-ROM.

Hirose, S. and Minematsu, N. (2002). 'Prosodic focus control in reply speech generation for a spoken #Dialogue System of Information Retrieval', in *Proceedings of the IEEE Workshop on Speech Synthesis*, Santa Monica: CD-ROM.

Hirschberg, J. (1991). 'Using text analysis to predict intonational boundaries. In *Proceedings of Eurospeech 91*, Geneva: 1275–1278.

Hirst, D. and Di Cristo, A. (1984). 'French intonation: a parametric approach', *Di Neueren Sprache* **83**: 554–569.

Holmes, J. (1983). 'Formant synthesizers: cascade or parallel?', *Journal of Speech Communication* **2**: 251–273.

Holmes, J. (1988). *Speech Synthesis and Recognition*. Wokingham: Van Nostrand Reinhold.

Holmes, J. and Holmes, W. (2001). *Speech Synthesis and Recognition*. London: Taylor & Francis.

Holmes, J., Mattingly, L. and Shearme, J. (1964). 'Speech synthesis by rule', *Language and Speech* **7**: 127–143.

Hyde, B. (2004). 'A restrictive theory of metrical stress', in *Phonology* **19**: 313–359.

Jacobson, R. (1960). 'Linguistics and poetics', in T. Sebeok (ed.), *Style in Language*. Cambridge, MA: MIT Press.

Jackendoff, R. (2002). *Foundations of Language: Brain, Meaning, Grammar, Evolution*. Oxford: Oxford University Press.

Jassem, W. (2002). 'Classification and organisation of data in intonation research', in A. Braun and H. Mastoff (eds), *Phonetics and its Applications: Festschrift for Jens-Peter Köster on the Occasion of his 60th Birthday*. Stuttgart: Franz Steiner Verlag, 289–297.

Johnson, D., Gardner, J. and Wiles, J. (2004). 'Experience as a moderator of the media equation: the impact of flattery and praise', *International Journal of Human–Computer Studies* **61**: 237–258.

Johnstone, T. and Scherer, K. (2000). 'Vocal communication of emotion', in M. Lewis and J. Haviland-Jones (eds), *Handbook of Emotions*. New York: Guilford Press, 220–235.

Johnstone, T., Van Reekum, C. and Scherer, K. (2001). 'Vocal expression correlates of appraisal processes', in K. Scherer, A. Schorr and T. Johnstone (eds), *Appraisal Processes in Emotion*. Oxford: Oxford University Press, 271–284.

Jones, D. (1950; 3rd edn 1967). *The Phoneme: its Nature and Use*. Cambridge: Heffer.

Kearns, K. (2000). *Semantics*. Basingstoke: Macmillan.

Keating, P. A. (1984). Phonetic and phonological representation of stop consonant voicing. *Language* **60**: 286–319.

Keating, P. A. (1990). Phonetic representations in a generative grammar. *Journal of Phonetics* **18**: 321–334.

Keller, E. (ed.) (1994). *Fundamentals of Speech Synthesis and Speech Recognition*. Chichester: Wiley.

Keller, E. (2002). 'Toward greater naturalness: future directions of research in speech synthesis', in E. Keller, G. Bailly, A. Monaghan, J. Terken and M. Huckvale (eds), *Improvements in Speech Synthesis*. Chichester: Wiley.

Kim, L. (2003). *The Official XMLSPY Handbook*. Indianapolis: Wiley.

Kirby, S. (1999). *Function, Selection, and Innateness: the Emergence of Language Universals*. Oxford: Oxford University Press.

Klabbers, E., van Santen, J. and Wouters, J. (2002). 'Prosodic factors for predicting local pitch shape', in *Proceedings of the IEEE Workshop on Speech Synthesis*, Santa Monica: CD-ROM.

Klatt, D. (1979). 'Synthesis by rule of segmental durations in English sentences', in B. Lindblom and S. Öhman (eds), *Frontiers of Speech communication*. New York: Academic Press, 287–299.

Klatt, D. (1980). 'Software for a cascade/parallel formant synthesizer', *Journal of the Acoustical Society of America* **67**: 971–995.

Klatt, D. and Klatt, L. (1990). 'Analysis, synthesis, and perception of voice quality variations among female and male talkers', *Journal of the Acoustical Society of America* **87**: 820–857.

Ladd, R. (1996). *Intonational Phonology*. Cambridge: Cambridge University Press.

Ladefoged, P. (1965). *The Nature of General Phonetic Theories*. Georgetown University Monograph on Languages and Linguistics, No. 18. Washington, DC: Georgetown University.

Ladefoged, P. (1996). *Elements of Acoustic Phonetics*. Chicago: University of Chicago Press.

Ladefoged, P. (2001). *Vowels and Consonants*. Oxford: Blackwell.

Lakoff, G. (1987). *Women, Fire, and Dangerous Things: What Categories Reveal About the Mind*. Chicago: University of Chicago Press.

Lawrence, W. (1953). 'The synthesis of speech from signals which have a low information rate', in W. Jackson (ed.), *Communication Theory*. New York: Butterworth.

Lazarus, R. (2001). 'Relational meaning and discrete emotions', in K. Scherer, A. Schorr and T. Johnstone (eds), *Appraisal Processes in Emotion*. Oxford: Oxford University Press, 37–69.

LeDoux, J. (1996). *The Emotional Brain*. New York: Simon & Schuster.

LeDoux, J. (2000) 'Cognitive–emotional interactions: listen to the brain', in R. Lane and L. Nadel (eds), *Cognitive Neuroscience of Emotion*. Oxford: Oxford University Press.

Lehiste, I. (1970). *Suprasegmentals*. Cambridge, MA: MIT Press.

Levinson, S. (1985). *Pragmatics*. Cambridge: Cambridge University Press.

Liu, C. and Kewley-Port, D. (2004). 'Vowel format discrimination for high-fidelity speech', *Journal of the Acoustical Society of America* **116**: 1224–1234.

Lindblom, B. (1983). 'Economy of speech gestures', in P. MacNeilage (ed.), *The Production of Speech*. New York: Springer-Verlag, 217–246.

Lindblom, B. (1990). 'Explaining phonetic variation: a sketch of the H and H theory', in W. Hardcastle and A. Marchal (eds), *Speech Production and Speech Modelling*. Dordrecht: Kluwer, 403–439.

Lindblom, B. (1991). 'Speech transforms on the extent, systematic nature, and functional significance of phonetic variation', *Speech Communication* **11**: 357–368.

Lisker, L. and Abramson, A. (1964). 'A cross-language study of voicing in initial stops: acoustical measurements', *Word* **20**(3).

Lubker, J. and Parris, P. (1970). 'Simultaneous measurements of intraoral air pressure, force of labial contact, and labial electromyographic activity during production of the stop consonant cognates /p/ and /b/', *Journal of the Acoustical Society of America* **47**: 625–633.

Luck, M. McBurney, P. and Preist, C. (2004). 'A manifesto for agent technology: towards next-generation computing', *Autonomous Agents and Multi-Agent Systems* **9**: 203–252.

MacNeilage, P. (1963). 'Electromyographic and acoustic study of the production of certain final clusters', *Journal of the Acoustical Society of America* **35**: 461–463.

MacNeilage, P. (1970). 'Motor control of serial ordering of speech', *Psychological Review* **77**: 182–196.

MacNeilage, P. and De Clerk, J. (1969). 'On the motor control of coarticulation in CVC monosyllables', *Journal of the Acoustical Society of America* **45**: 1217–1233.

Mermelstein, P. (1973). Articulatory model for the study of speech production. *Journal of the Acoustical Society of America* **53**: 1070–1082.

Milner, R. (1989). *Communication and Concurrency*. New Jersey: Prentice Hall.

Monaghan, A. (2002). 'State-of-the-art summary of European synthetic prosody R&D', in E. Keller, G. Bailly, A. Monaghan, J. Terken and M. Huckvale (eds), *Improvements in Speech Synthesis*. Chichester: Wiley, 93–103.

Morton, K. (1986). 'Cognitive phonetics: some of the evidence', in R. Channon and L. Shockey (eds), *In Honor of Ilse Lehiste*. Dordrecht: Foris, 191–194.

Morton, K. (1992). 'Pragmatic phonetics', in W. Ainsworth (ed.), *Advances in Speech, Hearing and Language Processing*, Vol. 2. London: JAI Press, 17–53.

Morton, K. and Tatham, M. (1980). 'Production instructions', in *Occasional Papers* 1980, Department of Language and Linguistics, University of Essex, 104–116.

Morton, K., Tatham, M. and Lewis, E. (1999). 'A new intonation model for text-to-speech synthesis', in J. Ohala (ed.), *Proceedings of the 14th International Congress of Phonetic Sciences*. Berkeley: University of California, 85–88.

Murray, I. and Arnott, J. (1993). 'Toward the simulation of emotion in synthetic speech: a review of the literature on human vocal emotion', *Journal of the Acoustical Society of America* **93**: 1097–1108.

Niedenthal, P., Halberstadt, J. and Innes-Ker, A. (1999). 'Emotional response categorization', *Psychological Review* **106**(2): 337–361.

Niedenthal, P., Auxiette, C., Nugier, A., Dalle, N., Bonin, P., Fayol, M. (2004). 'A prototype analysis of the French category "émotion"', *Cognition and Emotion* **18**(3): 289–312.

Oatley, K. and Johnson-Laird, P. (1987). 'Toward a cognitive theory of emotion', *Cognition and Emotion* **1**: 29–50.

Öhman, A., Flykt, A. and Esteves, F. (2001). 'Emotion drives attention: detecting the snake in the grass', *Journal of Experimental Psychology: General* **130**: 466–478.

Öhman, S. (1966). 'Coarticulation in VCV utterances: spectrographic measurements', *Journal of the Acoustical Society of America* **39**: 151–168.

Ortony, A., Clore, G. and Collins, A. (1988). *The Cognitive Structure of Emotions*. Cambridge: Cambridge University Press.

Panksepp, J. (1998). *Affective Neuroscience: the Foundations of Human and Animal Emotions*. Oxford: Oxford University Press.

Panksepp, J. (2000). 'Emotions as natural kinds within the mammalian brain', in M. Lewis and J. Haviland-Jones (eds), *Handbook of Emotions*. New York: Guilford Press, 137–156.

Pierrehumbert, J. (1981). 'Synthesizing intonation', *Journal of the Acoustical Society of America* **70**: 985–995.

Plutchik, R. (1994). *The Psychology and Biology of Emotion*. New York: HarperCollins.

Port, R., Cummins, F. and Gasser, M. (1996). 'A dynamic approach to rhythm in language: toward a temporal phonology', in B. Luka and B. Need (eds), *Proceedings of the Chicago Linguistic Society* **31**: 375–397.

Rolls, E. (1999). *The Brain and Emotion*. Oxford: Oxford University Press.

Pruden, R., d'Alessandro, C. and Boula de Mareüil, P. (2002). 'Prosidy synthesis by unit selection and transplantation of diphones', in *Proceeedings of the IEEEE Workshop on Speech Synthesis*, Santa Monica: CD-ROM.

Rubin, P. E., Baer, T. and Mermelstein, P. (1981). An articulatory synthesizer for perceptual research. *Journal of the Acoustical Society of America* **70**: 321–328.

SABLE Consortium (1998). Draft specification for Sable v.0.2. www.cstr.ed.ac.uk/projects/sable/sable_spec2.html.

Sharma, C. and Kunins, J. (2002). *Voice XML: Strategies and Techniques for Effective Voice Application Development with Voice XML 2.0*. New York: Wiley.

Scherer, K. (1993). 'Neuroscience projections to current debates in emotion psychology', *Cognition and Emotion* **7**: 1–41.

Scherer, K. (1996). 'Adding the affective dimension: a new look in speech analysis and synthesis', *Proceedings of the International Conference on Spoken Language Processing*. Philadelphia, 1014–1017.

Scherer, K. (2001). 'The nature and study of appraisal: a review of the issues', in K. Scherer, A. Schorr and T. Johnstone (eds), *Appraisal Processes in Emotion*. Oxford: Oxford University Press, 369–391.

Silverman, K., Beckman, M., Pitrelli, J., Ostendorff, M., Wrightman, C., Price, P., Pierrehumbert, J. and Hirschberg, J. (1992). 'ToBI: a standard for labelling English prosody', *in Proceedings of the 2nd International Conference on Spoken Language Processing (ICLSP)*, **2**: 867–870.

Sluijter, A., Bosgoed, E., Kerkhoff, J., Meier, E., Rietveld, T., Swerts, M. and Terken J. (1998). 'Evaluation of speech synthesis systems for Dutch in telecommunication applications', *Proceedings of the 3rd ESCA/COCOSDA Workshop of Speech Synthesis*. Jenolan Caves, Australia: CD-ROM.

Stevens, K. (2002). 'Toward formant synthesis with articulatory controls', in *Proceedings of the IEEE Workshop on Speech Synthesis*, Santa Monica: CD-ROM.

Stevens, K. and Bickley, C. (1991). 'Constraints among parameters simplify control of Klatt Formant Synthesizer', *Journal of Phonetics* **19**: 161–174.

Sunderland, R., Damper, R. and Crowder, R. (2004). 'Flexible XML-based configuration of physical simulations', *Software: Practice and Experience* **34**: 1149–1155.

Tams, A. (2003). 'Modelling intonation of read-aloud speaking styles for speech synthesis', unpublished PhD thesis. Colchester: University of Essex.

Tatham, M. (1970a). 'Articulatory speech synthesis by rule: implementation of a theory of speech production', *Report CN-534.1*. Washington: National Science Foundation. Also in Working Papers, *Computer and Information Science Research Center*, Ohio State University (1970).

Tatham, M. (1970b). 'Speech synthesis: a critical review of the state of the art', *International Journal of Man–Machine Studies* **2**: 303–308.

Tatham, M. (1971). 'Classifying allophones', *Language and Speech* **14**: 140–145.

Tatham, M. (1986a). 'Cognitive phonetics: some of the theory', in R. Channon and L. Shockey (eds), *In Honor of Ilse Lehiste*. Dordrecht: Foris, 271–276.

Tatham, M. (1986b). 'Towards a cognitive phonetics', *Journal of Phonetics* **12**: 37–47.

Tatham, M. (1995). 'The supervision of speech production', in C. Sorin, J. Mariani, H. Meloni and J. Schoentgen (eds), *Levels in Speech Communication: Relations and Interactions*. Amsterdam: Elsevier, 115–125.

Tatham, M. and Lewis, E. (1992). 'Prosodic assignment in SPRUCE text-to-speech synthesis', *Proceedings of the UK Institute of Acoustics* **14**: 447–454.

Tatham, M. and Lewis, E. (1999). 'Syllable reconstruction in concatenated waveform speech synthesis', *Proceedings of the International Congress of Phonetic Sciences*. San Francisco: 2303–2306.

Tatham, M. and Morton, K. (1969). 'Some electromyography data towards a model of speech production', *Language and Speech* **12**(1).

Tatham, M. and Morton, K. (1972). 'Electromyographic and intraoral air pressure studies of bilabial stops', in *Occasional Papers 12*. Colchester: University of Essex, 1–22.

Tatham, M. and Morton, K. (1980). 'Precision', in *Occasional Papers 23*. Colchester: University of Essex, 107–116.

Tatham, M. and Morton, K. (2002). 'Computational modelling of speech production: English rhythm', in A. Braun and H. R. Masthoff (eds), *Phonetics and Its Applications: Festschrift for Jens-Peter Köster on the Occasion of his 60th Birthday*. Stuttgart: Franz Steiner Verlag, 383–405.

Tatham, M. and Morton, K. (2003). 'Data structures in speech production', *Journal of the International Phonetic Association* **33**: 17–49.

Tatham, M. and Morton, K. (2004). *Expression in Speech: Analysis and Synthesis*. Oxford: Oxford University Press.

Tatham, M., Lewis, E. and Morton, K. (1998). 'Assignment of intonation in a high-level speech synthesiser', *Proceedings of the Institute of Acoustics* **20**: 255–262

Tatham, M., Morton, K. and Lewis, E. (2000). 'SPRUCE: speech synthesis for dialogue systems', in M. M. Taylor, F. Néel and D. G. Bouwhuis (eds), *The Structure of Multimodal Dialogue*, Vol. II. Amsterdam: John Benjamins, 271–292.

Taylor, P. (1995). 'The rise/fall/connection model of intonation', *Speech Communication* **15**: 169–186.

Taylor, P. (2000). 'Analysis and synthesis of intonation using the Tilt model', in *Journal of the Acoustical Society of America* **107**: 1697–1714.

Taylor, P., Black, A. and Caley, R. (1998). 'The architecture of the Festival speech synthesis system', in *Proceedings of the 3rd ESCA/COCOSDA Workshop of Speech Synthesis*, Jenolan Caves, Australia, 147–152: CD-ROM.

't Hart, J. and Collier, R. (1975). 'Integrating different levels of intonation analysis', *Journal of Phonetics* **3**: 235–255.

Tomkins, S. (1984). 'Affect theory', in K. Scherer and P. Eckman (eds), *Approaches to Emotion*. Hillsdale, NJ: Erlbaum, 163–195.

Traber. C. (1993). 'Syntactic processing and prosody control in the SVOX TTS system for German', *Proceedings of Eurospeech 93*. Berlin: 2097–2102.

Van Lancker-Sidtis, D. and Rallon, G. (2004). 'Tracking the incidence of formulaic expressions in everyday speech: methods for classification and verification', *Language and Communication* **24**: 207–240.

Verschueren, J. (2003). *Understanding Pragmatics*. London: Arnold.

W3C Consortium (2000). *XML Specification*. www.w3.org/XML/ .

W3C Consortium (2003). *SSML Specification*. www.w3.org/TR/2003/CR-speech-synthesis-20031218 .

W3C Consortium (2004). *VoiceXML Specification*. www.w3.org/TR/2004/REC-voicexml20-20040316/ .

Wang, W. S-Y. and Fillmore, C. (1961). 'Intrinsic cues and consonant perception', *Journal of Speech and Hearing Research* **4**: 130.

Wehrle, T. and Scherer, K. (2001). 'Toward computational modeling of appraisal theories', in K. Scherer, A. Schorr and T. Johnstone (eds), *Appraisal Processes in Emotion*. Oxford: Oxford University Press, 350–368.

Wells. J. (1982). *Accents of English*. Cambridge: Cambridge University Press.

Werner, E. and Haggard, M. (1969). 'Articulatory synthesis by rule', *Speech Synthesis and Perception: Progress Report 1*, Psychological Laboratory, University of Cambridge.

Wickelgren, W. (1969). 'Context-sensitive coding, associative memory and serial order in (speech) behavior', *Psychological Review* **76**: 1–15.

Wightman, C. and Ostendorf, M. (1994). 'Automatic labeling of prosodic patterns', *IEEE Transactions in Speech and Audio Processing* **2**: 469–481.

Young, S. (1999). 'Acoustic modelling for large-vocabulary continuous speech recognition', in K. Ponting (ed.), *Proceedings of the NATO Advanced Study Institute*. Springer-Verlag: 18–38.

Young, S. (2000). 'Probabilistic methods in spoken dialogue systems', in *Philosophical Transactions of the Royal Society* (Series A) 358 (1769): 1389–1402.

Young, S. (2002). 'Talking to machines (statistically speaking)', in *International Conference on Spoken Language Processing*, Denver: CD-ROM.

Young, S. and Fallside, F. (1979). 'Speech synthesis from concept: a method for speech output for information systems', *Journal of the Acoustical Society of America* **67**: 685–695.

Zajonc, R. (1980). 'Feeling and thinking: preferences need no inferences', *American Psychologist* **35**: 151–175.

Zellner, B. (1994). 'Pauses and the temporal structure of speech', in E. Keller (ed.), *Fundamentals of Speech Synthesis and Speech Recognition*. Chichester: Wiley, 41–62.

Zellner, B. (1998). 'Temporal structures for fast and slow speech rate', in *Proceedings of the 3rd ESCA/COCOSDA Workshop of Speech Synthesis*, Jenolan Caves, Australia: CD-ROM.

Zellner-Keller, B. and Keller, E. (2002). 'A nonlinear rhythmic component in various styles of speech', in E. Keller, G. Bailly, A. Monaghan, J. Terken and M. Huckvale (eds), *Improvements in Speech Synthesis*. Chichester: Wiley.

Author Index

Developments in Speech Synthesis   Mark Tatham and Katherine Morton
© 2005 John Wiley & Sons, Ltd.   ISBN: 0-470-85538-X

Index